ANALYSIS AND DESIGN

OF HIGH-VOLTAGE

TRANSMISSION

LINES

Jesus C. de Sosa

ANALYSIS AND DESIGN OF HIGH-VOLTAGE TRANSMISSION LINES

iUniverse books may be ordered through booksellers or by contacting:

iUniverse
1663 Liberty Drive
Bloomington, IN 47403
www.iuniverse.com
1-800-Authors (1-800-288-4677)

Because of the dynamic nature of the Internet, any Web addresses or links contained in this book may have changed since publication and may no longer be valid. The views expressed in this work are solely those of the author and do not necessarily reflect the views of the publisher, and the publisher hereby disclaims any responsibility for them.

ISBN: 978-1-4502-6613-0 (sc)
ISBN: 978-1-4502-6614-7 (ebook)

Printed in the United States of America

iUniverse rev. date: 10/12/2010

For my wife and children

TABLE OF CONTENTS

CHAPTER 7 SHUNT CAPACITANCE OF A TRANSMISSION LINE

CHAPTER 8 MODELS OF HIGH-VOLTAGE TRANSMISSION LINES

CHAPTER 9 ITERATIVE APPROACH IN SOLVING TRANSMISSION LINE PROBLEMS

CHAPTER 10   FINANCING A POWER SYSTEM

CHAPTER 11   A DESIGN PROJECT

REFERENCES

LIST OF TABLES

LIST OF FIGURES

PREFACE

A high-voltage power transmission system is built to match the needs of a population, businesses, and industries. Unfortunately, a given location may not have the resources to build a power plant. In this case, it must import power from other location. This book shows the essentials of high-voltage transmission lines. They are the medium for transferring power.

Chapter 1 is simplified introduction to high-voltage transmission line network. It represents the lines, loads, and internal impedances as resistances. Doing so prevents the introduction of complex numbers, which may complicate the goal of the analysis; namely, get an intuitive feel of how power flows from one bus to another.

Chapters 2 and 3 show examples of a two-bus network and two-generator in a two-bus network. Here, complex numbers are used in representing impedances and power. It introduces the basics of how power flow from one bus to another. The chapters also analyze the roles of voltage magnitudes and angles in controlling the power flow.

Chapter 4 introduces the concept of bus power as the difference between the generated power and the power demand. As will be seen in the latter chapters, the representation is an efficient one since it simplifies the mathematical representation of power flow in a transmission line.

Nowadays, conserving energy is at everyone's mind. Savings in energy can be realized by improving the power factor of a power system. Chapter 5 examines the use of shunt capacitors in minimizing the series impedance of a transmission line. It saves energy.

Chapters 6 and 7 examine the methods of calculating the inductance and capacitance of transmission lines. They are useful not only in characterizing a transmission line but also in design.

In chapter 8, three models of transmission lines are shown. They are the short line, medium length, and long-line models. The chapter also shows how the impedance and admittance of a line can be measured.

Chapter 9 is the iterative approach of solving high-voltage transmission line network. Using the Gauss iterative technique, it shows how the voltage of a bus and its current flow can be calculated.

Like any other investment, investing in a power plant requires careful financing. Chapter 10 explores capital cost and the operating cost as the two most important variables in successful financing. The chapter also introduces optimum strategy or optimum power dispatch to minimize the power generation of a plant.

Chapter 11 is an example of a design project. It encapsulates not only the technical requirements of a design but also its financial requirements. The project shows that the design of a high-voltage power transmission system is iterative not only on its technical requirements but also in minimizing its costs.

CHAPTER 1

RESISTIVE REPRESENTATION OF

IMPEDANCES IN A POWER SYSTEM

Using resistance, instead of impedance, to represent the circuit parameters of the sources, loads, and transmission lines makes the analysis of a power system easier. It provides an intuitive feel on how currents flow in the network. Additionally, it can easily show the limitations of uncompensated transmission lines in satisfying the maximum power transfer theorem.

1.1 Review of the Basic Tools of Circuit Analysis

This section summarizes the basic tools of circuit analysis that are so important in simplifying high-voltage transmission lines. The tools are current and voltage divisions, source transformation, superposition theorem, and the maximum power transfer theorem. Another important principle, although not emphasized in basic courses of circuit analysis, is the calculation of the resistance of a load from its power rating.

1.1.1 Current division

Suppose a current source delivers power to two resistors as shown on Figure 1.1. Then the current in each resistor are

$$I_1 = \left(\frac{R_2}{R_1 + R_2} \right) I$$

and

$$I_2 = \left(\frac{R_1}{R_1 + R_2} \right) I.$$

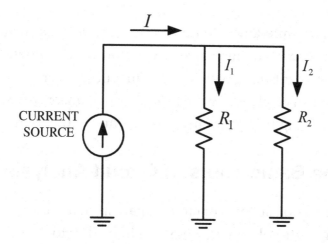

Figure 1.1 Circuit Illustrating Current Division

1.1.2 Voltage division

Voltage division is another useful in circuit analysis. For the circuit of Figure 1.2, the voltages across the resistors are given by

$$V_1 = \left(\frac{R_1}{R_1 + R_2} \right) V,$$

and

$$V_2 = \left(\frac{R_2}{R_1 + R_2} \right) V.$$

$$V_1 = \left(\frac{R_1}{R_1 + R_2} \right) V \qquad V_2 = \left(\frac{R_2}{R_1 + R_2} \right) V$$

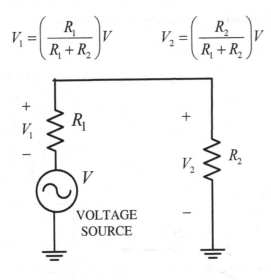

Figure 1.2 Circuit Illustrating Voltage Division

1.1.3 Source transformation

Source transformation states that a voltage source can be transformed to a current source by dividing the voltage by its internal resistance. The (parallel) internal resistance of the current source must be the same as the internal (series) resistance of the voltage source. Figure 1.3 illustrates a circuit with a voltage source and its equivalent with a current source. On the figure, calculations are also shown for the voltage across the second resistor of both circuits. Similar calculations can be made for the voltage across the first resistor.

Conversely, a current source with parallel resistance can be converted to a voltage source by multiplying the current source by its parallel resistance and inserting the resistance in series with the source.

$$V_2 = \left(\frac{R_1}{R_1 + R_2} \right)(I)(R_2)$$

$$= \left(\frac{R_1}{R_1 + R_2} \right)\left(\frac{V}{R_1} \right)(R_2)$$

$$= \left(\frac{R_2}{R_1 + R_2} \right)V$$

$$V_2 = \left(\frac{R_2}{R_1 + R_2} \right)V$$

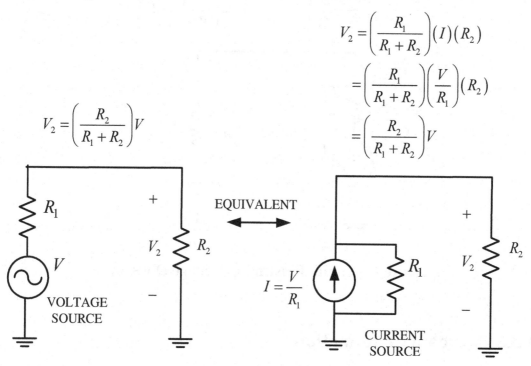

Figure 1.3 Circuit Illustrating Source Transformation

1.1.4 Maximum power transfer theorem

Consider a source with internal resistance and external load as shown on Figure 1.4. The power in the load resistor R_2 is

$$P_2 = I^2 R_2 = \left(\frac{V}{R_1 + R_2}\right)^2 (R_2) = \frac{V^2}{R_1^2 + 2R_1 R_2 + R_2^2}(R_2) = \frac{V^2}{\dfrac{R_1^2}{R_2} + 2R_1 + R_2}.$$

Taking the derivative of the equation with respect to R_2 and equating the result to zero gives $R_1 = R_2$. The theorem states that for the theorem to hold, the resistance of the source must be equal to the resistance of the load.

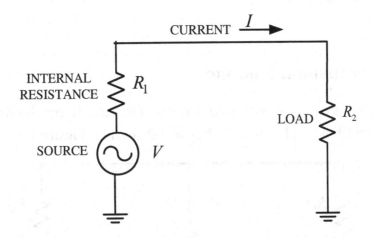

Figure 1.4 Circuit Illustrating Maximum Power Transfer Theorem

1.1.4.1 Modified form of maximum power transfer theorem

In some applications, the resistance of the source may be considered to be the equivalent resistance of series and parallel combinations of resistors. The load may also be considered to the equivalent resistance of several series and parallel resistors.

In order to compare the two equivalent resistances, it is more useful to represent the maximum power transfer theorem in terms of current sources. Figure 1.1 is the current-source equivalent of Figure 1.4. Immediately, and without additional calculations, the condition of maximum power occurs when the two resistances are equal. That is,

$$R_2 = R_1 .$$

The resistors R_1 and R_2 can be equivalent resistances of two or more resistors in series or parallel.

1.1.5 Superposition theorem

Figure 1.5 shows a circuit with two sources. The circuit may be replaced by two circuits with a single source. They are shown on Figure 1.6 (a) and (b).

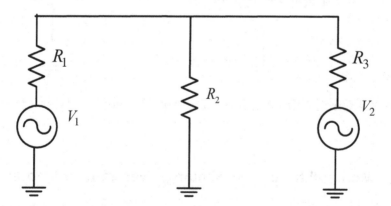

Figure 1.5 A Circuit with Two Sources

Superposition theorem allows the replacement of a circuit with N sources by N circuits with one source. The process is by killing all the sources except for one. A voltage source is killed by shorting its output terminals (making its voltage zero). Conversely, a current source is killed by opening one of its output terminals (stops the flow of current).

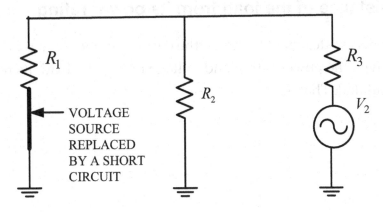

(a) First voltage source replaced by short circuit

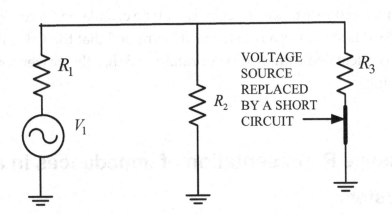

(b) Second voltage source replaced by short circuit

Figure 1.6 The Equivalent Circuits of the Single Circuit

1.1.6 Resistance of the load from its power rating

Power systems are designed to match the current and voltage requirements of a load. Given the power rating and voltage rating of a load, its resistance may be calculated. That is,

$$P = VI \qquad (1)$$

and

$$V = IR \qquad (2)$$

are two simultaneous equations.

There are two unknowns in the system of two equations above. They are the current, I and resistance, R. They can all be calculated (given P and V). When a load is shown as a resistance, it is implied that the value of the resistance was calculated from its power rating and that the power source can match the rating.

1.2 Resistive Representation of Impedances in a Power System

Figure 1.7 shows a power system consisting of two power plants, A and B, transmission lines, and loads.

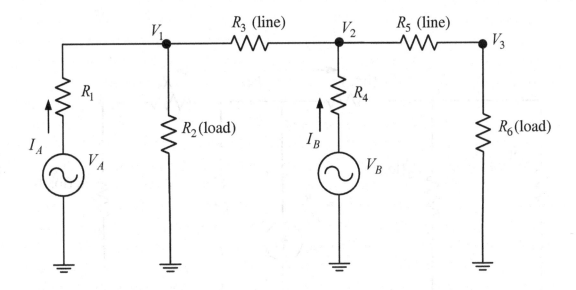

Figure 1.7 Two Power Plants with Resistive Loads and Transmission Lines

Power plant A has voltage V_A and current I_A. Similar notation applies to power plant B. Internal resistances are R_1 and R_4. Transmission line resistances are R_3 and R_5. The resistance of the load closest to the power plant A is R_2. Resistance of the load closest to the power plant B is R_6.

In the following developments, the superposition theorem, source conversion theorem, and current division will used in finding the total current flow from the plants.

1.2.1 Current flow when power plant A is "killed"

Figure 1.8 shows the case when the power plant A is "killed". It also shows the transformation of power plant B from voltage source to current source.

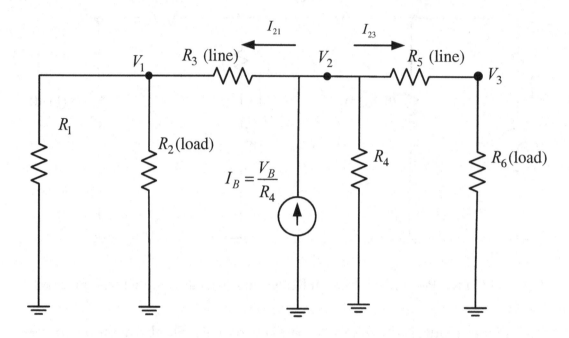

Figure 1.8 Current Flow when Power Plant A is "Killed"

The equivalent resistance of the fourth, fifth, and sixth resistors, as seen from node 2 to ground, is

$$R_{456} = R_4 \, / / \left(R_5 + R_6 \right).$$

Resistors 1, 2, and 3 have equivalent resistance of

$$R_{123} = \left(R_1 \, / / R_2 \right) + R_3.$$

Using current division, the branch currents to the equivalent resistances are:

$$I_{123} = \frac{R_{456}}{R_{456} + R_{123}} \left(\frac{V_B}{R_4} \right)$$

and

$$I_{456} = \frac{R_{123}}{R_{456} + R_{123}} \left(\frac{V_B}{R_4} \right).$$

These currents will be analyzed later.

1.2.2 Current flow when power plant B is "killed"

Figure 1.9 is the case when the power plant B is "killed". The equivalent resistance of resistors 3, 4, 5, and 6 is

$$R_{3456} = R_3 + R_4 // (R_5 + R_6).$$

Both resistors 1 and 2 are connected to the ground. Their equivalent resistance is

$$R_{12} = R_1 // R_2.$$

Using current division again,

$$I_{12} = \frac{R_{3456}}{R_{3456} + R_{12}} \left(\frac{V_A}{R_1} \right)$$

and

$$I_{3456} = \frac{R_{12}}{R_{3456} + R_{12}} \left(\frac{V_A}{R_1} \right).$$

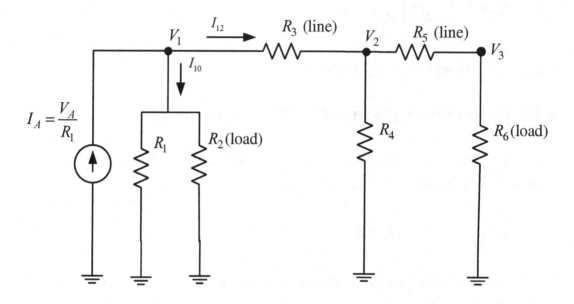

Figure 1.9 Current Flow when Power Plant B is "Killed"

1.2.3 The difficulty of matching the internal resistance of a power plant

Consider the first current calculated above. That is,

$$I_{123} = \left(\frac{R_{456}}{R_{123} + R_{456}} \right) \frac{V_B}{R_4}.$$

Rearranging the factors,

$$I_{123} = \left(\frac{R_{456}}{R_4}\right)\frac{V_B}{(R_{123}+R_{456})}.$$

The ideal value of $\dfrac{R_{456}}{R_4}$ is 1.0 since it will not scale down V_B. In order for the ratio to be 1.0,

$$R_4 = R_{456} = R_4 // (R_5 + R_6).$$

Solving the equation gives

$$\frac{\dfrac{R_4(R_5+R_6)}{R_4+(R_5+R_6)}}{R_4} = 1.0.$$

Simplifying,

$$\frac{(R_5+R_6)}{R_4+(R_5+R_6)} = 1.0.$$

The only possible value R_4 is zero which is an impossible case.

The next current is

$$I_{456} = \frac{R_{123}}{R_{456}+R_{123}}\left(\frac{V_B}{R_4}\right).$$

With this current, the ideal case should be

$$R_4 = R_{123} = (R_1 // R_2) + R_3.$$

Since the resistance of the transmission line, R_3 is greater than the internal resistance, R_4 of the plant, then it is impossible to satisfy the equation. That is,

$$R_4 < (R_1 // R_2) + R_3$$

must be true. All that can be done is to make the resistance of the transmission as small as possible.

Performing the same analysis for

$$I_{12} = \frac{R_{3456}}{R_{3456} + R_{12}} \left(\frac{V_A}{R_1} \right)$$

and

$$I_{3456} = \frac{R_{12}}{R_{3456} + R_{12}} \left(\frac{V_A}{R_1} \right)$$

gives

$$R_1 = 0 \text{ (an impossible case),}$$

and

$$R_1 = R_{3456} = R_3 + R_4 // (R_5 + R_6).$$

Again, since $R_3 > R_1$, it is impossible to satisfy the equality. The best that can be done is to make the resistance of the transmission line R_3 as small as possible. The ideal value of the resistance of the transmission line, R_5 should be as small as possible.

14

In the above analysis, the currents $I_{123} = \dfrac{R_{456}}{R_{456} + R_{123}}\left(\dfrac{V_B}{R_4}\right)$ and

$I_{3456} = \dfrac{R_{12}}{R_{3456} + R_{12}}\left(\dfrac{V_A}{R_1}\right)$ have opposite directions in the common transmission

line, R_3. When both sources are on, their difference determines the net current flowing in a direction. It is this net current that is responsible for the power flow between nodes (or busses) 1 and 2. In the next chapters, such power flow will be examined closely.

1.2.4 Application of the maximum power transfer theorem to equivalent resistances

While it may be impossible to make the resistance of a load, seen by a source, equal to the source's internal resistance, it is possible to make the equivalent resistances (of several resistors) equal. As shown in paragraph 1.1.4.1, the form of circuit with current sources is especially suited for this case. Specifically, in the two-power plant example above, the power plant that was "killed" produces two equivalent resistances that are parallel with each other. From their form, the equality can be formed.

As an example, when the power plant A was killed, two equivalent resistances were formed. Figure 1.10 shows them. When applied to the circuit, the maximum power transfer theorem requires that

$R_{456} = R_{123}$.

There are six resistors in the circuit. The resistances of the two transmission lines are R_3 and R_5. Since the internal resistances of the source and the loads can't be adjusted, only the resistances of the transmission lines can be modified. To minimize transmission line losses, the equivalent resistance (R_{456}, or R_{123}) with the smaller value must be used as the reference. The higher equivalent resistance must be modified to equal the lower equivalent resistance.

When power plant B is killed, the theorem requires satisfying the following equivalent resistances:

$$R_{3456} = R_{12}.$$

The equation shows that the transmission line resistances, R_3 and R_5 are in the left-hand side. Since both are on the same side, it may be impossible to satisfy the equality. That is,

$$R_{3456} = R_3 + R_4 // (R_5 + R_6)$$

has to be greater than $R_{12} = R_1 // R_2$. In such a case, the maximum power transfer theorem is not be realizable. Again, as in the previous section, the best that can be done is to make the resistances of the transmission lines as small as possible.

In chapter 5, shunt capacitance will be used in minimizing the resistance of a line. The main benefit of the above analysis shows how a shunt capacitance can affect the equivalent resistance of a group of resistances.

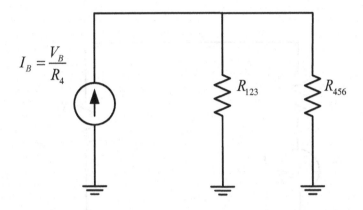

THE MAXIMUM POWER TRANSFER
THEOREM REQUIRES

$$R_{456} = R_{123}$$

Figure 1.10 Application of the Maximum Power Transfer Theorem to
Equivalent Resistances when Power Plant A is "Killed"

1.3 The Rise in Voltage of a Bus when the Number of its Generators Increases

The following circuit analysis addresses the problem of voltage rise in a bus
when the number of generators connected to it increases. It is important in
synchronization since when the amount of rise is larger than the upper limit,
it may not be able to synchronize.

Consider the circuit shown on Figure 1.11. By voltage division, the
voltage across the load, R_2 is

$$V_2 = \left(\frac{R_2}{R_1 + R_2} \right) V .$$

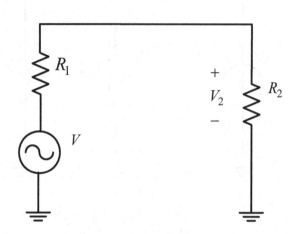

Figure 1.11 One Generator Connected to a Bus

Suppose the number of generators connected to the bus increases twice (see Figure 1.12).

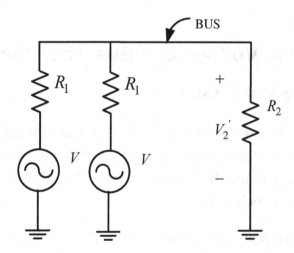

Figure 1.12 Two Generators Connected to a Bus

Converting the voltage sources into current sources gives the equivalent circuit of Figure 1.13.

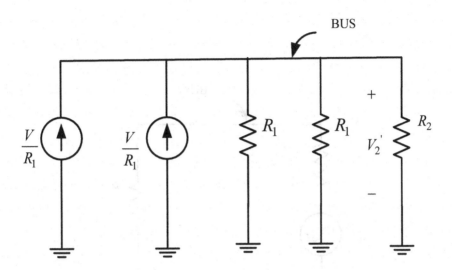

Figure 1.13 Equivalent Circuit using Current Sources

Adding the two current sources and getting the equivalent resistance of the two R_1's gives another equivalent circuit of Figure 1.14. Performing another source transformation, from current to voltage source, gives the final equivalent circuit of Figure 1.15.

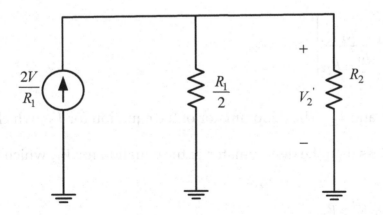

Figure 1.14 Another Equivalent Circuit

19

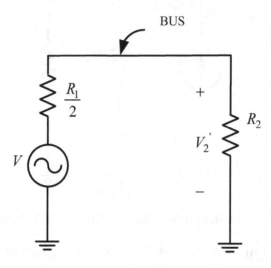

Figure 1.15 Final Equivalent Circuit

Applying voltage division, the voltage in the bus (across R_2) is

$$V_2' = \left(\frac{R_2}{\dfrac{R_1}{2} + R_2} \right) V \, .$$

Compare V_2 and V_2'. The denominator of the equation for V_2', which is

$\dfrac{R_2}{\dfrac{R_1}{2} + R_2}$, is less than the denominator of the equation for V_2, which is $R_1 + R_2$.

It shows that $V_2' > V_2$.

CHAPTER 2

POWER FLOW IN A TWO-BUS SYSTEM

The previous chapter represents elements in a power system as resistor. This chapter will start using impedance throughout the remaining part of the book to achieve the required accuracy starting with power flow analysis. Power flow analysis is a steady state analysis. It analyzes the response of a power system long time after switching occurs. Two variables are critical in power flow analysis. They are voltage magnitude and power angle. Before analyzing such variables, review of electrical quantities as complex number and per unit value is appropriate.

2.1   Review of Complex Number

In electrical engineering, variables such as voltage, current, impedance, and power are all complex numbers. In order to conserve notations, the following conventions will be used.

The real part of a variable will be designated as "Re" and its imaginary part by "Im" (after the "j" complex number operator). As an example, voltage is understood to be

$$V = \text{Re}(V) + j\,\text{Im}(V)$$

where

$\text{Re}(V)$ = real part of the voltage V, and

$\text{Im}(V)$ = imaginary part of the voltage V.

For current, the convention will be

$$I = \text{Re}(I) + j\,\text{Im}(I).$$

While impedance can be represented by $Z = \text{Re}(Z) + j\,\text{Im}(Z)$ it is more compact to represent the same by

$$Z = R + jX$$

where

R = resistance or real part of Z, and

X = reactance or imaginary part of Z.

Similarly, power will be represented by

$$S = P + jQ$$

where

P = real part of S, and

Q = imaginary part of S.

2.1.1 Conjugate operation

The conjugate of a complex number is the same number except that its imaginary part is its opposite or conjugate. An asterisk, as a superscript, will indicate the conjugate of a variable. For voltage, the conjugate of

$$V = \text{Re}(V) + j\,\text{Im}(V)$$

is

$$V^* = \mathrm{Re}(V) - j\,\mathrm{Im}(V).$$

For power, the conjugate of

$$S = P + jQ$$

is

$$S^* = P - jQ.$$

2.1.2 Polar form of complex number

The above representation of complex number is in rectangular form of the Cartesian coordinate system. Division and multiplication of complex numbers are easier if they are in polar form.

In general, the polar form of

$$x = a + jb$$

is

$$x = \left| \sqrt{a^2 + b^2} \right| \angle \theta$$

where

a = real part of x,

b = imaginary part of x, and

$$\theta = \tan^{-1}\left(\frac{b}{a}\right).$$

Figure 2.1 shows the relationship between the rectangular and polar forms of a complex number.

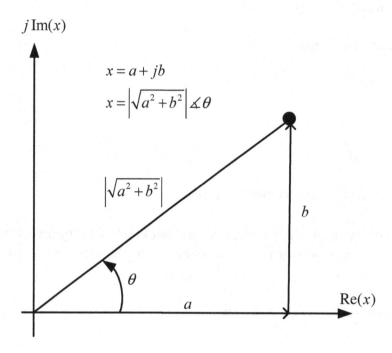

Figure 2.1 Rectangular and Polar Forms of a Complex Number

2.1.3 Per unit value of electrical quantities with transformers

Impedances of transformers are expressed in percent or per unit. The impedance is for the whole transformer itself and independent of whether it is "seen" at the primary or secondary side.

In order to avoid re-calculating the impedances of the lines, as reflected on the primary or secondary side, the impedances of bus-to-bus sections of power lines are calculated in percent or per unit also.

The first step in calculating per unit impedance consists of finding the base power, base voltages, base currents, and the base impedances on the primary secondary sides of a transformer. They are shown below:

$$\left| S_{B,1} \right| = \left| \frac{S_{3\phi}}{3} \right|,$$

$$V_{B,1} = V_{LN,1} = \frac{V_{LL,1}}{\sqrt{3}},$$

$$I_{B,1} = \frac{\left| S_{B,1} \right|}{V_{B,1}}, \text{ and}$$

$$Z_{B,1} = \frac{V_{B,1}}{I_{B,1}}.$$

where

$S_{B,1}$ = base power in the primary side of the power transformer,

$S_{3\phi}$ = the rated three-phase power of the power transformer,

$V_{B,1}$ = base voltage in the primary side of the power transformer,

$V_{LN,1}$ = line-to-neutral voltage across the primary side of the power transformer,

$V_{LL,1}$ = line-to-line voltage across the primary side of the power transformer, and

$I_{B,1}$ = base current in the primary branch of the power transformer.

The same calculations are made on the secondary side of the transformer.

In summary, the base calculations require a system of four independent equations with six unknowns. To find the solution of the system, two of the unknowns should be known or given. These unknowns are, usually, the three-phase rated power of the transformer and its line-to-line rated voltages.

2.1.4 Per unit value of electrical quantities without transformers

In case the section of a transmission line has no transformer, the rated line-to-line voltage of a bus may be used as the base voltage provided the base power is a 3-phase power and impedance is a 3-phase impedance.

2.2 Real and Reactive Power Flows in a Two-Bus System

Figure 2.2 is a two-bus power system. A transmission line connects the two busses. In the figure, it is assumed that the bus voltages are controlled by power plants delivering power to the bus or loads consuming power from it.

The derivation of the real power flow and the reactive power flow starts with

$$S_{ij} = V_i I^*$$

where

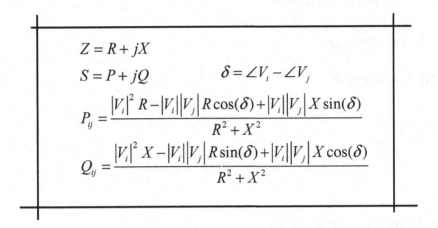

$$Z = R + jX$$

$$S = P + jQ \qquad \delta = \angle V_i - \angle V_j$$

$$P_{ij} = \frac{|V_i|^2 R - |V_i||V_j|R\cos(\delta) + |V_i||V_j|X\sin(\delta)}{R^2 + X^2}$$

$$Q_{ij} = \frac{|V_i|^2 X - |V_i||V_j|R\sin(\delta) + |V_i||V_j|X\cos(\delta)}{R^2 + X^2}$$

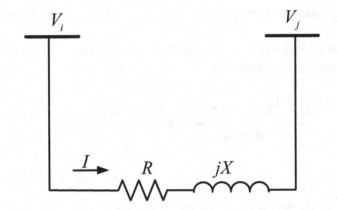

Figure 2.2 A Two-Bus System Separated by a Transmission Line

S_{ij} = complex power from bus i to bus j,

V_i = voltage at bus I, and

I^* = complex current.

The complex current is

$$I^* = \frac{V_i^* - V_j^*}{Z^*}$$

where

V_j = voltage at bus j, and

Z = impedance of the transmission line between the busses.

After several more definitions and simplifications, the real and reactive power flows from bus i to bus j are

$$P_{ij} = \frac{|V_i|^2 R - |V_i||V_j| R\cos(\delta) + |V_i||V_j| X \sin(\delta)}{R^2 + X^2}$$

and

$$Q_{ij} = \frac{|V_i|^2 X - |V_i||V_j| R\sin(\delta) + |V_i||V_j| X \cos(\delta)}{R^2 + X^2}$$

where

P_{ij} real power flow from bus i to bus j,

Q_{ij} = reactive power flow from bus i to bus j,

R = resistance of the line, and

X = reactance of the line.

In the above equations, the power angle is the difference in the phase angles of the two voltages. That is,

$$\delta = \angle V_i - \angle V_j.$$

Power flow is not a power loss. The former is a transmission of power. It depends on both the resistive and the inductive reactance of impedance.

Real power loss depends on the resistive component only. It is zero when the resistance of the line is zero. However, real power flow exists even if the resistance is zero since real power flow also depends with the inductive reactance of the transmission line. The similarly can be extended to reactive power loss and reactive power flow with one difference. In transmission line, the inductive reactance is always greater than the resistance. Consequently, zero reactive power loss is meaningless.

2.3 Control of Power Flow by Voltage Magnitudes and Power Angle

Consider the two busses shown on Figure 2.2. The data on the busses are as follows:

$$V_i = 230 KV \angle 10^0$$

$$V_j = 240 KV \angle 0^0$$

$$Z = 0.02 + j0.15 \, pu$$

$$|S_B| = 200MVA$$

(a) Determine the real power flow, the reactive power flow, the real power loss, the real reactive loss in per unit values.

(b) Assume that $V_j = 230KV\angle 0^0$. Repeat part (a) above.

(c) Assume that $V_i = 230KV\angle 15^0$. Repeat part (a) above.

(d) Tabulate and analyze the results of parts (a), (b), and (c).

Solution:

Part (a). The following are the calculated per unit value of the voltages:

$$V_{i,pu} = \frac{230KV\angle 10^0}{230KV} = 1.0\angle 10^0$$

$$V_{j,pu} = \frac{240KV\angle 0^0}{230KV} = 1.043\angle 0^0 .$$

The real power flow equation

$$P_{ij,pu} = \frac{|V_{i,pu}|^2 R_{pu} - |V_{i,pu}||V_{j,pu}|R_{pu}\cos(\delta) + |V_{i,pu}||V_{j,pu}|X_{pu}\sin(\delta)}{R_{pu}^2 + X_{pu}^2}$$

gives

$$P_{ij,pu} = \frac{|1.0|^2 (0.02) - |1.0||1.043|(0.02)\cos(10^0) + |1.0||1.043|(0.15)\sin(10^0)}{(0.02)^2 + (0.15)^2}$$

or,

$$P_{ij,pu} = 1.16\,pu(watt).$$

The corresponding per unit reactive power flow is

$$Q_{ij,pu} = \frac{\left|V_{i,pu}\right|^2 X_{pu} - \left|V_{i,pu}\right|\left|V_{j,pu}\right| R_{pu}\sin(\delta) - \left|V_{i,pu}\right|\left|V_{j,pu}\right| X_{pu}\cos(\delta)}{R_{pu}^2 + X_{pu}^2}.$$

After substituting the corresponding values,

$$Q_{ij,pu} = -0.336\,pu(var).$$

The per unit current is

$$I_{ij,pu} = \frac{V_{i,pu} - V_{j,pu}}{R_{pu} + jX_{pu}}.$$

Substituting the values,

$$I_{ij,pu} = \frac{1.0\angle 10^0 - 1.043\angle 0^0}{0.02 + j0.15}.$$

The polar form of the complex number must be converted into its rectangular form since addition and subtraction are not allowed operations in polar form. The result is

$$I_{ij,pu} = \frac{0.98 + j0.17 - (1.04 + j0)}{0.02 + j0.15} = \frac{-0.06 + j0.17}{0.02 + j0.15}.$$

For division, it is easier to convert the rectangular form to its polar form. This gives

31

$$I_{ij,pu} = \frac{0.18\angle109.44^0}{0.15\angle82.41^0} = 1.20\angle27.03\,pu(amp).$$

The real power loss in the line is

$$P_{loss,pu} = II^*R = I^2R = (1.20)^2(0.02) = 0.03\,pu(watt).$$

The reactive power loss in the line is

$$Q_{loss,pu} = II^*R = I^2X = (1.20)^2(0.15) = 0.22\,pu(var).$$

<u>Part (b).</u> Using $V_j = 230KV\angle0^0$ and repeating the same calculations as the above give:

$$P_{ij,pu} = 1.148\,pu(watt)$$

$$Q_{ij,pu} = -0.048\,pu(var)$$

$$I_{ij,pu} = 1.152\angle23.6^0$$

$$P_{loss,pu} = 0.0265\,pu(watt)$$

$$Q_{loss,pu} = 0.199\,pu(var).$$

<u>Part (c).</u> Using $V_i = 230KV\angle15^0$ for bus i,

$$P_{ij,pu} = 1.764\,pu(watt)$$

$$Q_{ij,pu} = -0.284\,pu(var)$$

$$I_{ij,pu} = 1.725\angle25.9^0$$

$$P_{loss,pu} = 0.0595\,pu(watt)$$

$$Q_{loss,pu} = 0.4463\,pu(var).$$

Part (d). Table 2.1 summarizes the results of parts (a), (b), and (c). It shows the variation of real power flow and reactive power flow as a function of voltage magnitudes and power angle. Note that the impedance of the transmission line is constant in all parts.

Table 2.1 Power Flow Analysis: Summary of Calculations

Part	V_i	V_j	P_{ij}	Q_{ij}	I_{ij}	P_{loss}	Q_{loss}
(a)	$230\angle10^0$	$240\angle0^0$	1.160	-0.336	$1.20\angle27.03^0$	0.030	0.220
(b)	$230\angle10^0$	$230\angle0^0$	1.148	-0.048	$1.152\angle23.6^0$	0.026	0.199
(c)	$230\angle15^0$	$240\angle0^0$	1.764	-0.284	$1.725\angle25.9^0$	0.059	0.446

2.3.1 Analysis of real power flow

Table 2.2 replicates the first four columns of Table 2.1. It shows that the flow of real power, P_{ij}, from bus i to bus j, is always positive. To understand such a result, examine the formula for the real power flow.

The flow of real power is positive if

$$0 < \frac{|V_i|^2 R - |V_i||V_j| R\cos(\delta) + |V_i||V_j| X \sin(\delta)}{R^2 + X^2}$$

Simplifying further,

$$|V_j| R \cos(\delta) + |V_j| X \sin(\delta) < |V_i| R.$$

Table 2.2 Variation of Real Power Flow

Part	V_i	V_j	P_{ij}
(a)	$230 \angle 10^0$	$240 \angle 0^0$	1.160
(b)	$230 \angle 10^0$	$230 \angle 0^0$	1.148
(c)	$230 \angle 15^0$	$240 \angle 0^0$	1.764

The above inequality has five unknowns. To simplify the analysis further, use ratios to reduce the number of unknowns. Dividing both sides of the inequality by the first term, gives the simpler result

$$1 + \left(\frac{X}{R}\right) \tan(\delta) < \left(\frac{|V_i|}{|V_j|}\right)\left(\frac{1}{\cos(\delta)}\right).$$

Alternatively, the condition when the real power flow is greater than zero is

$$\left(\frac{|V_i|}{|V_j|}\right)\left(\frac{1}{\cos(\delta)}\right) - \left(\frac{X}{R}\right)\tan(\delta) > 1. \qquad \text{(condition for } P_{ij} > 0\text{)}$$

The last inequality reduces the number of unknowns to two ratios and the power angle. This is the last reducible form.

Since $\tan(\delta) > 1$ when $\delta > 45^0$, the inequality may not be satisfied if the power angle is greater than 45 degrees. However, a power angle greater than zero is favorable since $\dfrac{1}{\cos(\delta)}$ becomes larger. In Table 2.2, the power angle for part (c) is 15 degrees. It is greater than the 10 degrees power angle of part (a).

Parts (a) and (c) have also the same voltage magnitudes. Their effect on the real power flow should be the same. Part (b) has ratio of one for their voltage magnitudes. Its power angle, which is 10 degrees, is less than the 15-degree power angle of part (c). The greater power angle of part (c) made its real power flow greater than past (b).

In summary, for given $\left(\dfrac{X}{R}\right)$ ratio, the real power flow becomes more positive if:

- The power angle is greater than zero but less than 45 degrees, and
- The ratio of their voltage magnitudes should be close to one.

2.3.2 Analysis of reactive power flow

Table 2.3 is the variation of the reactive power flow as the magnitude of the voltages and their power angle. A reactive power flow is negative or less than zero if

$$0 > \frac{|V_i|^2 X - |V_i||V_j| R\sin(\delta) + |V_i||V_j| X\cos(\delta)}{R^2 + X^2}.$$

After some more algebra, the condition for the reactive power flow to be negative is to satisfy the inequality

$$\left(\frac{R}{X}\right)\tan(\delta)-1>\left(\frac{|V_i|}{|V_j|}\right)\left(\frac{1}{\cos(\delta)}\right).$$

Alternatively, the condition of negative reactive power flow occurs when

$$\left(\frac{R}{X}\right)\tan(\delta)-\left(\frac{|V_i|}{|V_j|}\right)\left(\frac{1}{\cos(\delta)}\right)>1. \quad \text{(condition for } Q_{ij}<0)$$

Table 2.3 Variation of Reactive Power Flow

Part	V_i	V_j	Q_{ij}
(a)	$230\angle10^0$	$240\angle0^0$	-0.336
(b)	$230\angle10^0$	$230\angle0^0$	-0.048
(c)	$230\angle15^0$	$240\angle0^0$	-0.284

Parts (a) and (c) show the condition when the ratio of the voltage magnitudes is less than one. Hence, the two parts have a more negative reactive power flow when compared with part (b).

The second term of the left-hand side of the inequality must be as small as possible. This means that $\cos(\delta)$ must be large as possible. Since the maximum possible value of $\cos(\delta)$ is one, when δ is zero, then the power angle must be as small as possible. In Table 2.3, part (a) has smaller power angle than part (c). Therefore, par (a) must be more negative.

36

In summary, for given $\left(\dfrac{X}{R}\right)$ ratio, the reactive power flow becomes more negative if:

- The ratio of voltage magnitudes is less than one, and
- The power angle must be small as possible.

2.3.3 Analysis of real and reactive power losses

The goal of transmission line design is to minimize its power losses. As shown on Table 2.4, part (b) has the least real and reactive power losses. It is the best. Part (b) has voltage magnitudes that are equal with each other. Its power angle is relatively low.

The next best case is part (a). Its real and reactive power losses are less than part (c).

Table 2.4 Variation in the Real and Reactive Power Losses

Part	V_i	V_j	P_{loss}	Q_{loss}
(a)	$230\angle10^0$	$240\angle0^0$	0.030	0.220
(b)	$230\angle10^0$	$230\angle0^0$	0.026	0.199
(c)	$230\angle15^0$	$240\angle0^0$	0.059	0.446

CHAPTER 3

TWO-BUS TWO-GENERATOR SYSTEM

The previous example shows two busses with voltages. A more realistic example consists of busses with generators and loads.

3.1 A Two-Bus, Two-Generator Problem

Consider the two-bus power system shown on Figure 3.1. Bus 2 is more efficient and operating at full capacity. The following are the data on the power system:

$V_1 = V_2 = 138\,\text{KV}$

$P_{G1}(\text{max}) = 160\,\text{MW}$ (less efficient – allowed to swing)

$S_{D1} = 20 + j10$

$Z = j0.062$

$P_{G2} = 28\,\text{MW}$ (more efficient and operating at full capacity of 28 MW)

$S_{D2} = 50 + j30$

$|S_b| = 200\,\text{MVA}$ (base power)

The problem is to find the complex power that each generator must produce.

Because the generator at bus 2 is more efficient, it must be operating

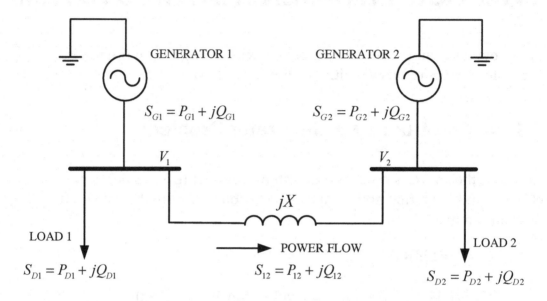

GENERATOR 2 IS ASSUMED MORE EFFICIENT.
THE PROBLEM IS TO FIND THE COMPLEX
POWER EACH GENERATOR MUST GENERATE.

Figure 3.1 A Case of Two-Bus, Two-Generator Problem

at full capacity. Note that in the analysis, the maximum power at bus 1 is used. Summing the real powers in bus 2,

$$P_{G2} = P_{D2} + P_{21}.$$

Solving for P_{21}

$$P_{21} = P_{G2} - P_{D2} = 28 - 50 = -22 \text{ MW}.$$

The flow of real power from bus 1 to bus 2 is the opposite of P_{21}. That is,

$$P_{12} = 22 \text{ MW}.$$

Summing the real powers in bus 1,

$$P_{G1} = P_{D1} + P_{12} = 20 + 22 = 42 \text{ MW.}$$

Similarly, the sum of the real power powers in bus 2 must be the generator power in the same bus. That is,

The next step is to find the power angle. This, however, will require per unit values. In per unit,

$$P_{12} = \frac{22 \text{ MW}}{200 \text{ MVA}} = 0.11 \text{ pu watt.}$$

Substituting values in

$$P_{12} = \frac{|V_1||V_2|\sin(\delta)}{X}$$

$$0.11 = \frac{(1.0)(1.0)\sin(\delta)}{0.062}$$

gives

$$\delta = 0.398^0.$$

Since,

$$Q_{12} = \frac{|V_1|^2 - |V_1||V_2|\cos(\delta)}{X}$$

$$Q_{12} = \frac{(1.0)^2 - (1.0)(1.0)\cos(0.398^0)}{0.062} = (389.1)(10^{-6}) \text{ pu var.}$$

The per unit value of the reactive load at bus 1 is

$$Q_{D1} = \frac{10 \text{ MVAR}}{200 \text{ MVA}} = 0.05 \text{ pu var.}$$

Now, the total reactive power that generator 1 must generate is

$$Q_{G1} = Q_{12} + Q_{D1} = (389.1)(10^{-6}) + 0.05 = 0.0503 \text{ pu var.}$$

To find the reactive power that generator 2 must produce, first find the reactive power from bus 2 to bus 1. It is

$$Q_{21} = \frac{(1.0)^2 - (1.0)(1.0)\cos(0.398^0)}{0.062} = (389.1)(10^{-6}) \text{ pu var.}$$

The per unit value of the reactive load at bus 2 is

$$Q_{D2} = \frac{30 \text{ MVAR}}{200 \text{ MVA}} = 0.15 \text{ pu var.}$$

Generator 2 must produce the reactive power of

$$Q_{G2} = Q_{21} + Q_{D2} = (389.1)(10^{-6}) + 0.15 = 0.1504 \text{ pu var.}$$

Converting the per unit values to actual values,

$$Q_{G1} = 0.0503(200 \text{ MVA}) = 10 \text{ MVAR}$$

$$Q_{G2} = 0.1504(200 \text{ MVA}) = 30.$$

The complex power of each generator is:

$$S_{G1} = P_{G1} + jQ_{G1} = 42 \text{ MW} + j10 \text{ MVAR}$$

$$S_{G2} = P_{G2} + jQ_{G2} = 28 \text{ MW} + j30 \text{ MVAR}.$$

Compare the above values with the original generating capacity of each generator. These capacities are:

$$P_{G1}(\max) = 160 \text{ MW (less efficient)}$$

$P_{G2} = 28$ MW (more efficient and operating fully at all times).

Generator 1 must generate 42 MW, which is less than the 160 MW. In contrast, the real power that generator 2 must produce at all times is 28 MW.

3.2 Systematic Analysis of the Calculations

The calculations above are lengthy. To develop the approach or strategy in the calculations, the flowing paragraphs highlight the important points in the calculations.

3.2.1 Calculations for the complex power of generator 1

Consider analyzing the set of equations that relates to solving the complex power that the first generator has to generate. The set has six equations:

$$S_{G1} = P_{G1} + jQ_{G1} \tag{1}$$

$$P_{G1} = P_{D1} + P_{12} \tag{2}$$

$$P_{12} = -\left(P_{G2} - P_{D2}\right) \tag{3}$$

$$P_{12} = \frac{|V_1||V_2|\sin(\delta)}{X} \tag{4}$$

$$Q_{12} = \frac{|V_1|^2 - |V_1||V_2|\cos(\delta)}{X} \tag{5}$$

$$Q_{G1} = Q_{D1} + Q_{12} \tag{6}$$

There are thirteen variables in the equations. Seven of the thirteen variables are given. They are:

- Two voltage magnitudes,

- A reactance,

- Two real power demands,

- One generated real power, and

- One reactive power demand.

The system of six equations can solve the remaining six unknown variables. These variables are S_{G1}, P_{G1}, Q_{G1}, P_{12}, δ, and Q_{12}.

In summary, the central problem in solving the unknowns is to solve for the power angle. Solution to finding the power angle must involve the two busses. Since generator 2 is more efficient and operating fully at all times, the difference between the real power it produces and its real power demand is the flow of power from bus 2 to bus 1. Taking the negative of this flow gives the flow of real power from bus 1 to bus 2. The power angle may be found from the latter power flow. Once the angle is known, the reactive power flow can be calculated.

3.2.2 Calculations for the complex power of generator 2

For the sake of completeness, the complex power that generator must generate satisfies the following equations

$$S_{G2} = P_{G2} + jQ_{G2} \tag{1}$$

$$Q_{G2} = Q_{D2} + Q_{21} \tag{2}$$

$$Q_{21} = \frac{|V_2|^2 - |V_1||V_2|\cos(\delta)}{X}. \tag{3}$$

The two voltage magnitudes and reactance are given. As for the power angle, it was calculated from the complex power that generator 1 must generate. There are two other constants. They are the generated real power, and reactive power demand in bus 2. The result is a system of three equations with three unknowns (S_{G2}, Q_{G2}, and Q_{21}).

3.2.3 Energy conservation

The above example illustrates how energy conservation. By operating the more efficient generator or plant at its full capacity, energy can be saved. As the load varies, the other generator or plant can deliver the additional power demand.

3.2.4 Remark on finding the power angle by calculations

The example above also shows how to find a power angle by calculations. If two busses are tens or hundreds of miles away, it is virtually impossible to measure the power angle. For accurate measurements, it will need a single oscilloscope with two channels. One of the channels will measure the first bus and the other the second bus. The two channels must measure at the same time. It means triggering both channels at the same time. With long separation between the busses, it may be impossible to do so.

CHAPTER 4

BUS POWER AND POWER FLOW

This chapter introduces bus power and its complex conjugate as the power in a transmission line. In addition to showing current flow in a transmission line with shunt capacitance, the chapter also shows approximation to the resistance of the line (using loss factor). The approximation will be used in expressing bus power in as a power flow involving circuit parameters such as impedance, voltage magnitudes, and power angles. Equations from the approximation provide additional helpful insights on how the circuit parameters affect complex power.

4.1 Bus Power

To simplify the analysis of power transmission systems, it will be desirable to subtract the power demand from the generated power. The difference is the bus power.

Consider a bus with a generator and a power demand directly connected with it. The bus power is the difference between the generated power and power demand. In general,

$$S_i = S_{Gi} - S_{Di}$$

where

S_i = power in bus i,

S_{Gi} = generated power in bus i, and

$$S_{Di} = \text{power demand at bus } i.$$

Since

$$S_{Gi} = P_{Gi} + jQ_{Gi}$$

and

$$S_{Di} = P_{Di} + jQ_{Di}$$

then

$$S_i = P_{Gi} + jQ_{Gi} - (P_{Di} + jQ_{Di})$$

or

$$S_i = (P_{Gi} - P_{Di}) + j(Q_{Gi} - Q_{Di}).$$

Essentially, the bus power is the difference between the generated power and the power demand. The symbol of a bus power is similar to an AND gate as shown on Figure 4.1.

4.1.1 Transmitted Power as Complex Conjugate of Bus Power

Figure 4.2 shows a simple circuit with source impedance Z_1 and load impedance Z_2. The complex power in Z_1 is

$$S_1 = II^*Z_1 = (IZ_1)I^* = (V - V_1)I^*.$$

For the source to deliver maximum power to the load, the constraint on the load must be

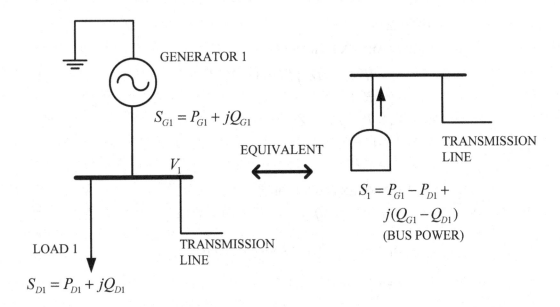

Figure 4.1 Illustration of Bus Power

$$Z_2 = Z_1^* = R_1 - jX_1.$$

Since Z_2 is the complex conjugate of Z_1, the voltage across Z_1 must be equal to the voltage across Z_2. That is,

$$V - V_1 = V_1.$$

Therefore,

$$S_1 = V_1 I^*.$$

With the above constraint, the complex power in Z_2 is

$$S_2 = II^* Z_2 = II^* Z_1^* = I\left(I^* Z_1^*\right) = S_1^*.$$

NOTES:

1. COMPLEX POWER OF Z_1

$$S_1 = II^* Z_1 = (IZ_1)I^* = (V - V_1)I^*$$
$$Z_1 = R_1 + jX_1$$
$$Z_2 = R_1 - jX_1 = Z_1^*$$
$$V - V_1 = V_1$$
$$S_1 = V_1 I^*$$

2. COMPLEX POWER OF Z_2

$$S_2 = II^* Z_2 = II^* Z_1^* = I(I^* Z_1^*) = S_1^*$$

$$S_1^* = V_1^* I \quad \text{Compare with} \quad S_1 = V_1 I^*$$

$$I = \frac{S_1^*}{V_1^*}$$

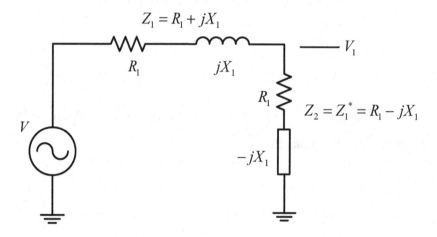

Figure 4.2 Circuit Illustrating Complex Conjugate of a Power Source

Note that S_2 is the complex conjugate S_1^*. Using the commutative property for multiplication

$$S_1^* = II^* Z_1^* = I\left(I^* Z_1^*\right) = I\left(V_1^*\right).$$

Finally,

$$I = \frac{S_1^*}{V_1^*}.$$

The last equation will be used in developing the current flow in the transmission line.

Note that putting a constraint on the load impedance as complex conjugate of the source impedance makes the circuit resistive. That's, the reactive components of the source and the load cancel each other out.

In terms of bus power and transmission line, if a bus has complex power, S_1 then the transmission line must have complex power, S_1^*. That is,

$$P_{Gi} - P_{Di} = \text{ real part of } S_i^*$$

and

$$Q_{Gi} - Q_{Di} = \text{ imaginarty part of } S_i^*$$

where

$P_{Gi} - P_{Di}$ = real part of the bus power,

$Q_{Gi} - Q_{Di}$ = imaginary part of the bus power, and

S_i^* = complex power of the transmission line.

Figure 4.3 is current-equivalent source of Figure 4.2. The figure shows the

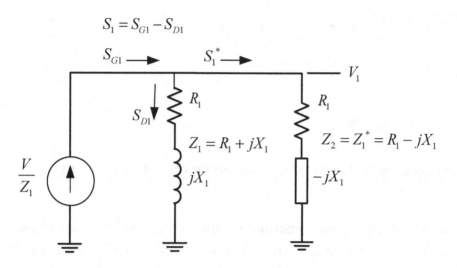

Figure 4.3 Equivalent Circuit of Figure 4.2 showing the Complex Powers

complex power of the source as S_1 and its complex conjugate, S_1^* in the load. In the flowing developments, the load is the transmission line.

4.2 Current Flow in a Transmission Line

Consider Figure 4.4. The current leaving bus 1 is

$$I_1 = \frac{S_1^*}{V_1^*}.$$

The current in the shunt capacitor, Y_1 is

$$I_{Y1} = V_1 Y_1.$$

Current in the series impedance is

$$I = \frac{(V_1 - V_2)}{Z}.$$

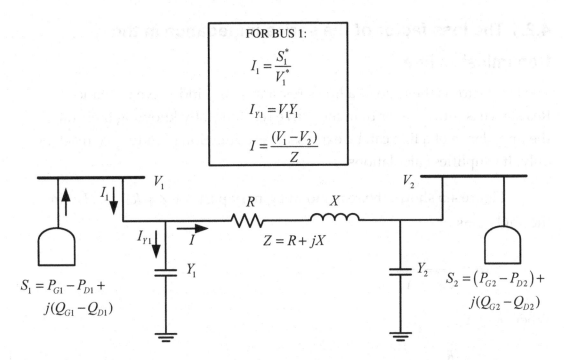

Figure 4.4 Currents in a Transmission Line with Shunt Capacitance

The current leaving bus 1 is the sum of the two currents

$$I_1 = I_{Y1} + I$$

or,

$$I_1 = V_1 Y_1 + \frac{(V_1 - V_2)}{Z}.$$

Except for the difference in subscripts, the current leaving bus 2 is

$$I_2 = V_2 Y_2 + \frac{(V_2 - V_1)}{Z}.$$

4.2.1 The loss factor of the series impedance in the transmission line

The loss factor is the ratio of a line's resistance and inductive reactance. Resistance is much lower than inductive reactance. By knowing their ratio, the impedance of a line can be expressed as a function of inductive reactance only. It simplifies calculations.

Figure 4.5 shows the real and imaginary parts of $Z = R + jX$. Define the angle α as

$$\sin \alpha = \frac{R}{\sqrt{R^2 + X^2}}$$

When $R \ll X$,

$$\sin \alpha \approx \frac{R}{X}.$$

For small angles,

$$\sin \alpha \approx \alpha$$

so that

$$\alpha \approx \frac{R}{X}.$$

Finally, the resistance can be expressed as

$$R \approx \alpha X.$$

Now, the series impedance may now be expressed as

$$Z \approx \alpha X + jX$$

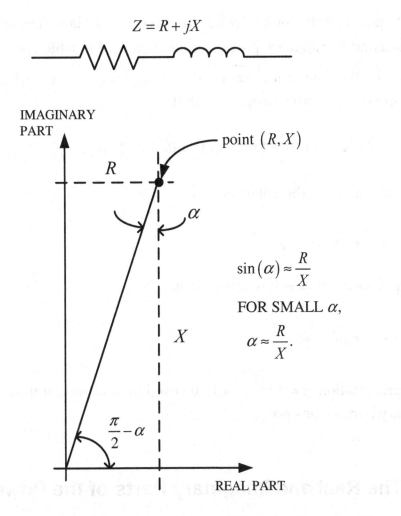

Figure 4.5 Real and Imaginary Parts of Z on the Complex Plane

or

$$Z = (\alpha + j)X.$$

The reactance X becomes a common factor of both the resistance and the reactance. Note that the line has always the resistance $R = \alpha X$ and not zero.

Figure 4.6 is the plot of $\alpha + j$ on the complex plane. The real part of $\alpha + j$ is α and it imaginary part is the coefficient of j, which is 1.0. For small values of α, the magnitude $\sqrt{\alpha^2 + 1.0^2}$ reduces to $\sqrt{1.0^2} = 1.0$. Therefore, the polar form of $\alpha + j$ can be approximated by

$$\alpha + j = \left|\sqrt{\alpha^2 + 1.0^2}\right| \angle (\pi/2 - \alpha) \approx 1.0 \angle (\pi/2 - \alpha).$$

In exponential form, the polar form is

$$1.0 \angle (\pi/2 - \alpha) = e^{j\left(\frac{\pi}{2} - \alpha\right)}.$$

The impedance can now be expressed as

$$Z = R + jX \approx Xe^{j\left(\frac{\pi}{2} - \alpha\right)}.$$

The approximation $Z \approx Xe^{j\left(\frac{\pi}{2} - \alpha\right)}$ will be used in interpreting the real and imaginary parts of bus power.

4.3   The Real and Imaginary Parts of the Power Flow

It will now be instructive to find, using loss factor, how shunt capacitance and series impedance of transmission lines affect power flow.

The reactance of a capacitor is given by

$$X_C = \frac{1}{\omega C}.$$

Its impedance is

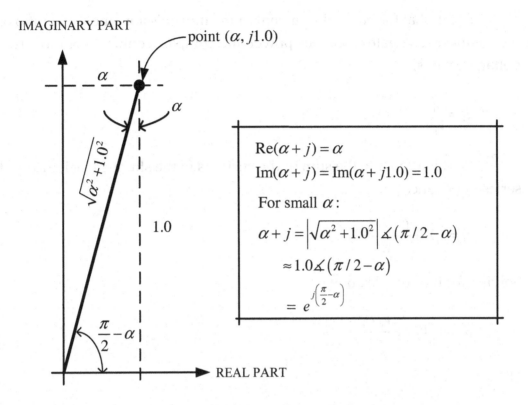

Figure 4.6 Plot of $\alpha + j$ on the Complex Plane

$$Z_C = \frac{1}{sC} = \frac{1}{j\omega C} = \frac{-j}{\omega C}.$$

In general, the inverse of impedance is admittance, or

$$Y_C = \frac{1}{Z_C}.$$

After some algebra,

$$Y_C = \frac{j}{X_C}.$$

57

Recall that for bus 1, the current in the transmission line is the ratio of the complex conjugate of the bus power and complex conjugate of the bus voltage. That is,

$$I_1 = \frac{S_1^*}{V_1^*}.$$

It is also equal to the sum of the currents in the shunt capacitance and series impedance

$$I_1 = V_1 Y_1 + \frac{(V_1 - V_2)}{Z}.$$

Solving for the complex power,

$$S_1^* = V_1^* V_1 Y_1 + \frac{V_1^*(V_1 - V_2)}{Z}.$$

Using

$$Z \approx X e^{j\left(\frac{\pi}{2} - \alpha\right)},$$

$$V_1 = |V_1| e^{j\delta_1},$$

$$V_2 = |V_2| e^{j\delta_2},$$

and

$$e^{j\theta} = \cos\theta + j\sin\theta,$$

the complex bus power $S_1^* = (P_{G1} - P_{D1}) - j(Q_{G1} - Q_{D1})$ becomes

$$S_1^* = (P_{G1} - P_{D1}) - j(Q_{G1} - Q_{D1}) = \frac{|V_1|^2}{X_C} \sin\alpha - \frac{|V_1||V_2|}{X} \sin[\alpha - (\delta_1 - \delta_2)]$$

$$-j\left\{\frac{|V_1|^2}{X_C} - \frac{|V_1|^2}{X} \cos\alpha + \frac{|V_1||V_2|}{X} \cos[\alpha - (\delta_1 - \delta_2)]\right\}.$$

Similarly, for the second bus

$$S_2^* = (P_{G2} - P_{D2}) - j(Q_{G2} - Q_{D2}) = \frac{|V_2|^2}{X_C} \sin\alpha - \frac{|V_1||V_2|}{X} \sin[\alpha + (\delta_1 - \delta_2)]$$

$$-j\left\{\frac{|V_2|^2}{X_C} - \frac{|V_2|^2}{X} \cos\alpha + \frac{|V_1||V_2|}{X} \cos[\alpha + (\delta_1 - \delta_2)]\right\}.$$

While the last two equations are long they are not complex. All the variables such as voltage magnitudes, inductive reactance, capacitive reactance, phase angles, and the loss factor can be measured or calculated. Therefore, the equations can be solved.

4.3.1 Comparison between the power flow equations and the bus power equations

It will be interesting to find out how the power flow equations compare with the bus power equations. The power flow from bus 1 to bus 2 is

$$S_{12} = P_{12} + jQ_{12}.$$

Its real part is

$$P_{12} = \frac{|V_1|^2 R - |V_1||V_2| R\cos(\delta) + |V_1||V_2| X \sin(\delta)}{R^2 + X^2}.$$

In the derivation of the loss factor, it is assumed that $R \ll X$. Suppose $R = 0$ in the power flow equation. Then, the real power flow reduces to

$$P_{12} = \frac{0 + |V_1||V_2| X \sin(\delta)}{0 + X^2} = \frac{|V_1||V_2| X \sin(\delta)}{X^2} = \frac{|V_1||V_2| \sin(\delta)}{X}.$$

Compare the last equation with the real power of bus 1

$$P_{G1} - P_{D1} = \frac{|V_1|^2}{X_C} \sin\alpha - \frac{|V_1||V_2|}{X} \sin[\alpha - (\delta_1 - \delta_2)].$$

Except for the first term in the latter equation, the two equations are equivalent. The first term (involving X_C) is absent from the real power equation because the shunt capacitance is neglected during the derivation of the power flow equation.

The same assumptions and derivation gives similar form for the reactive power flow and reactive bus power. In conclusion, the complex conjugate of the bus power is the power flow between the two busses.

4.3.2 Effect of the variables on the bus power

For bus 1, its bus power is the complex conjugate of S_1^*. That is,

$$S_1 = (P_{G1} - P_{D1}) + j(Q_{G1} - Q_{D1}).$$

Its real power part is

$$P_{G1} - P_{D1} = \frac{|V_1|^2}{X_C} \sin\alpha - \frac{|V_1||V_2|}{X} \sin[\alpha - (\delta_1 - \delta_2)]$$

and its imaginary part is

$$Q_{G1} - Q_{D1} = \frac{|V_1|^2}{X_C} - \frac{|V_1|^2}{X} \cos\alpha + \frac{|V_1||V_2|}{X} \cos[\alpha - (\delta_1 - \delta_2)].$$

There are six independent variables in each equation. They are shown on Table 4.1. Note that increasing the shunt capacitive reactance, which is equivalent to decreasing the capacitance itself, decreases both the real power and the imaginary power. It means bus 1 need not produce as much either power.

Conversely, increasing the loss factor increases the real power and the reactive power. As a result, the bus has to generate more power. Increasing loss factor means increasing resistance. Hence, as the resistance of the transmission line increases, its real power and reactive power increases.

Finally, increasing the difference of the loss factor and the power angle also decreases the real power and the reactive power.

Table 4.1 Effect of the Variables in the Bus Power

Increasing Variable	Effect on: $P_{G1} - P_{D1}$ (note 1)	Effect on: $Q_{G1} - Q_{D1}$ (note 2)
Shunt capacitive reactance, X_C (or decreasing the capacitance)	Decreases	Decreases
Series inductive reactance	Increases	Decreases
Voltage magnitude $\lvert V_1 \rvert$	Increases	Increases
Voltage magnitude $\lvert V_2 \rvert$	Decreases	Increases
Loss factor, α	Increases	Increases
Difference of the loss factor and power angle, $\alpha - (\delta_1 - \delta_2)$	Decreases	Decreases

Notes: 1. $\quad P_{G1} - P_{D1} = \dfrac{\lvert V_1 \rvert^2}{X_C} \sin \alpha - \dfrac{\lvert V_1 \rvert \lvert V_2 \rvert}{X} \sin[\alpha - (\delta_1 - \delta_2)]$

2. $\quad Q_{G1} - Q_{D1} = \dfrac{\lvert V_1 \rvert^2}{X_C} - \dfrac{\lvert V_1 \rvert^2}{X} \cos \alpha + \dfrac{\lvert V_1 \rvert \lvert V_2 \rvert}{X} \cos[\alpha - (\delta_1 - \delta_2)]$

CHAPTER 5

POWER FACTOR IMPROVEMENT

A shunt capacitive reactance can improve a transmission line by minimizing its impedance and creating a complex conjugate of the load. This chapter shows two approaches in determining the size of the shunt capacitive reactance. The first approach estimates the shunt capacitive reactance as equal to the inductive reactance of the line. In the second approach, the capacitive reactance is assumed to be a part of the load (assumed inductive as is the usual case). Complex conjugate operation is then performed on the resulting equivalent circuit. As will be seen, the second approach is more accurate and general. Finally, the chapter also shows how shunt capacitive reactance can improve the voltage regulation across the line and saves power.

5.1 Location of Capacitor Banks

Table 4.1 from the previous chapter shows the effect of shunt capacitive reactance near a bus that generates power. As the shunt capacitive reactance increases, the power flow decreases. Since capacitive reactance increases with decreasing capacitance, installing capacitor banks near a bus that generate power is not recommended.

The ideal location of capacitor banks are far away from a power plant or generator site. They are usually located at substations at the end of the transmission lines feeding distribution lines.

Capacitor banks are installed to minimize the inductive reactance of a transmission line. As a result, the voltage at the end of the transmission line will increase. Since current is proportional to the difference of the voltages across the ends of a transmission line, the smaller difference in the voltages is desirable. Lower current means lower power losses.

5.2   Power Factor Improvement

This section examines closely the industry approach of sizing shunt capacitive reactance to compensate series inductive reactance of a transmission line. While the approach works, it increases the real power loss of the transmission line. Against this background, an approach treating the shunt capacitive reactance as part of the (inductive) load is made. It turned out that the latter approach is more accurate and decreases the real and reactive power losses in a transmission line.

5.2.1 Shunt capacitive reactance as complex conjugate of the inductive reactance

Suppose a power source is connected to a load by a transmission line with impedance $R_1 + jX_1$ as shown on Figure 5.1. Typical load in power transmission system is also resistive and inductive. It can be represented by $Z_2 = R_2 + jX_2$. Obviously, the maximum power transfer theorem could not be satisfied by such a setup. That is, for maximum power transfer to occur, the load impedance must be the complex conjugate of the source impedance (in this case is the transmission line).

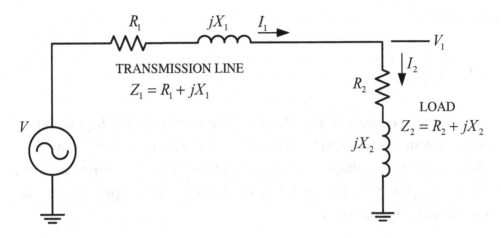

Figure 5.1 A Circuit with a Source, Transmission Line, and Load

Suppose a capacitor is installed after the transmission line and parallel with the load. See Figure 5.2.The size of the capacitor is chosen such that it is the conjugate of the inductive reactance in the transmission line and satisfies

$$\omega L_1 - \frac{1}{\omega C_1} = 0.$$

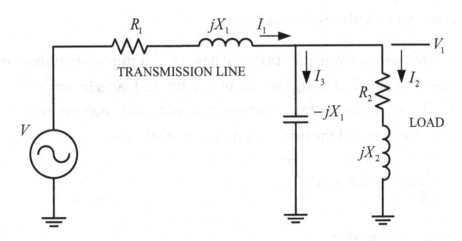

Figure 5.2 Shunt Capacitance at the End of a Transmission Line

Solving for C_1

$$C_1 = \left(\frac{1}{\omega^2}\right)\left(\frac{1}{L_1}\right).$$

To get the equivalent impedance of the resulting circuit (and verify if conjugate matching can occur), cut the load at the point indicated by the broken line, short the voltage source, and get the Thevenin impedance of the two parallel impedances. Figure 5.3 shows the steps of the procedure. The new equivalent impedance is

$$\left(R_1 + jX_1\right)//\left(-jX_1\right) = \frac{X_1^{\,2}}{R_1} - jX_1.$$

Figure 5.4 shows the equivalent circuit of the original circuit with the new equivalent impedance. Since the inductance of a line is usually higher than its resistance, the new equivalent impedance shows that the new resistance ($\frac{X_1^{\,2}}{R}$) has increased. Furthermore, the inductive reactance (jX_1) is replaced by capacitive reactance ($-jX_1$).

The new equivalent impedance has $-jX_1$. It implies that there is, although not perfect, conjugate matching with the load's inductive reactance of jX_2. To obtain the perfect conjugate matching, between the new transmission line and the load, it is necessary that:

$$\frac{X_1^{\,2}}{R_1} - jX_1 = R_2 + jX_2.$$

Equating the real parts,

NOTE:

$$Z_{eq} = \left(R_1 + jX_1\right)//\left(-jX_1\right) = \frac{X_1^2}{R} - jX_1$$

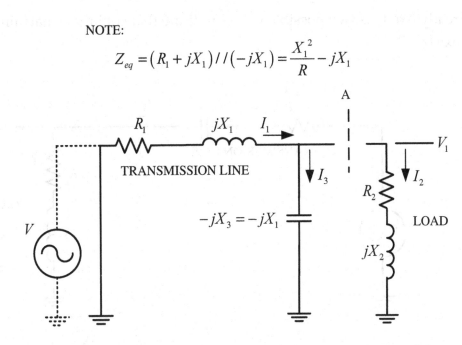

Figure 5.3 Steps in Finding the New Equivalent Impedance of the Transmission Line

$$\frac{X_1^2}{R_1} = R_2 ,$$

$$X_1^2 = R_1 R_2 , \text{ or}$$

$$X_1 = \left| \sqrt{R_1 R_2} \right| .$$

Equating the imaginary parts,

$$X_1 = X_2 .$$

The last two equations say that there are two possible values for X_1. Since X_1 is already fixed, its two possible values indicate that conjugate matching may not exists.

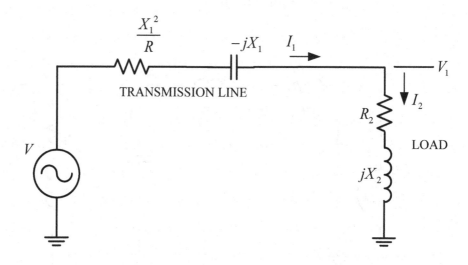

Figure 5.4 The New Equivalent Impedance of the Transmission Line with the Effect of the Shunt Capacitance

5.2.2 Shunt capacitive reactance with the effect of the load

Consider Figure 5.5. On the figure, a broken line encloses the shunt capacitance and load. Another broken line encloses the transmission line. In the approach, the equivalent impedance of the shunt capacitance and load is obtained. Its real part is then equated with the real part of the transmission line. Similarly, its imaginary part is equated with the imaginary part of the transmission line. Note that the voltage across the load is now designated V_{1N}

to indicate the new voltage. As will be seen, there will be an increase in the voltage across the load.

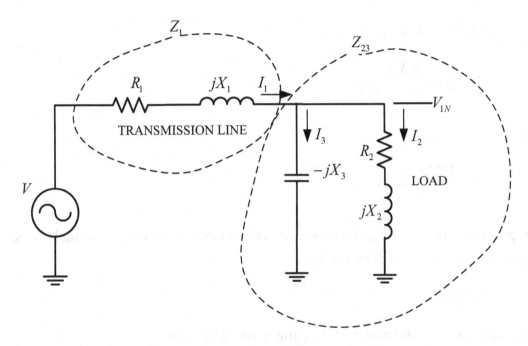

Figure 5.5 Shunt Capacitive Reactance Across the Transmission Line and Ground

The impedance of the transmission line is

$$Z_1 = R_1 + jX_1.$$

Z_{23}, the equivalent impedance of the circuit elements with subscripts 2 and 3, is

$$Z_{23} = \left(\frac{(R_2 + jX_2)(-jX_3)}{R_2 + jX_2 - jX_3} \right) = \frac{X_2X_3 - jR_2X_2}{R_2 + j(X_2 - X_3)}.$$

After rationalizing the denominator,

$$Z_{23} = \frac{R_2 X_3^2}{R_2^2 + (X_2 - X_3)^2} - j\frac{\left(R_2^2 X_3 + X_2^2 X_3 - X_2 X_3^2\right)}{R_2^2 + (X_2 - X_3)^2}.$$

For conjugate matching to occur,

$$R_1 = \frac{R_2 X_3^2}{R_2^2 + (X_2 - X_3)^2}$$

and

$$X_1 = \frac{\left(R_2^2 X_3 + X_2^2 X_3 - X_2 X_3^2\right)}{R_2^2 + (X_2 - X_3)^2}.$$

In practical applications, the resistance of the transmission line is much larger than the resistance of the load. Using

$$R_2 \approx 0,$$

the equations for the real and imaginary parts become

$$R_1 = 0$$

and

$$X_3 \approx \frac{X_1 X_2}{X_1 + X_2}.$$

The result shows that the capacitive reactance must be the parallel equivalent of the inductive reactance of the transmission line and the inductive reactance of the load.

The corresponding value of the capacitance can be found from

$$\frac{\omega L_1 L_2}{L_1 + L_2} - \frac{1}{\omega C_3} = 0.$$

Solving for the capacitance,

$$C_3 = \left(\frac{1}{\omega^2}\right)\left(\frac{L_1 + L_2}{L_1 L_2}\right) = \left(\frac{1}{\omega^2}\right)\left(\frac{1}{L_1} + \frac{1}{L_2}\right).$$

The value of the capacitance using this approach is larger than when simply assuming that the shunt capacitive reactance is equal to the inductive reactance of the line.

5.2.3 Verifying the real power loss and the reactive power loss

In section 1.2.1, the procedure for getting the equivalent impedance of the shunt capacitance in parallel with the impedance of the transmission line was show. The same procedure will be shown here except that the shunt capacitive reactance will no longer be equal to the inductive reactance of the transmission line. It will be X_3. Figure 5.6 shows the circuit.

Performing exactly the same procedure, the new equivalent impedance of the transmission line is

$$Z_{1N} = R_{1N} - jX_{1N}$$

where

$$R_{1N} = \frac{R_1 X_3^2}{R_1^2 + (X_1 - X_3)^2}, \text{ and}$$

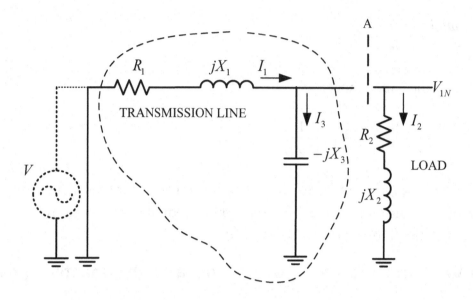

Figure 5.6 The Circuit with Shunt Capacitance X_3

$$X_{1N} = \frac{R_1^2 X_3 + X_1^2 X_3 - X_1 X_3^2}{R_1^2 + (X_1 - X_3)^2}.$$

Note that the values of R_1 and X_1 are from the original transmission line, X_2 is from the load, and X_3 is given by

$$X_3 \approx \frac{X_1 X_2}{X_1 + X_2}.$$

Furthermore, note the new voltage, V_{1N} across the load. Figure 5.7 shows the equivalent circuit.

It will be interesting to find out if R_{1N} is less than R_1 since doing so will prove that the new equivalent circuit consumes less real power loss. That is,

$$R_{1N} \leq ? R_1 .$$

Substituting the values,

$$\frac{R_1 X_3^2}{R_1^2 + (X_1 - X_3)^2} \leq ? R_1 .$$

After simplifying,

$$0 \leq ? R_1^2 + X_1^2 - 2X_1 X_3 .$$

Since

$$X_3 \leq X_1 ,$$

then

$$0 \leq R_1^2 + X_1^2 - 2X_1 X_3 .$$

Therefore, the new equivalent resistance is less than the original resistance and it consumes less power.

Doing the same for the reactive component, the following inequality should be verified

$$X_{1N} \leq ? X_1 .$$

That is,

$$\frac{R_1^2 X_3 + X_1^2 X_3 - X_1 X_3^2}{R_1^2 + (X_1 - X_3)^2} \leq ? X_1 .$$

There is no common factor for X_1 on the left hand side. Suppose $X_1 = X_3$. Then, the last inequality reduces to

<u>MATH SUMMARY:</u>

$$Z_{1N} = R_{1N} - jX_{1N}$$

$$X_{1N} = \frac{R_1^2 X_3 + X_1^2 X_3 - X_1 X_3^2}{R_1^2 + (X_1 - X_3)^2}$$

$$R_{1N} = \frac{R_1 X_3^2}{R_1^2 + (X_1 - X_3)^2}$$

$$X_3 \approx \frac{X_1 X_2}{X_1 + X_2}$$

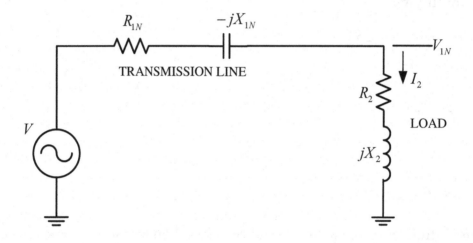

Figure 5.7 The Equivalent Impedance of the Transmission Line with the Shunt Capacitance

$$\frac{R_1^2 X_1}{R_1^2} \leq ? X_1 .$$

Finally,

$$1 \leq 1$$

and the inequality is satisfied.

Suppose $X_3 = 0$. Then,

$$\frac{0}{R_1^2 + X_1^2} \leq ? X_1$$

and

$0 \leq X_1$ (Yes).

Therefore, when the inequality is satisfied when $X_3 = X_1$ and $X_3 = 0$. Since the maximum possible value of X_3 is X_1, the inequality

$$X_{1N} \leq X_1$$

is satisfied at all times. It means the reactive power loss in the new equivalent impedance is less than that of the original.

5.3   Improvement in Voltage Regulation and Power Savings

Due to its lower impedance, there will be less voltage drop across the transmission line. Additionally, the load is conjugate matched to the line. The change in voltage can be expressed as

$$\frac{\Delta V}{V_1} = \left(\frac{V_{1N} - V_1}{V_1} \right)$$

where

ΔV = change in voltage,

V_{1N} = the new voltage in node 1, and

V_1 = original voltage in node 1.

Testing for the voltage rise simply involve measuring the voltage across the load before and after the shunt reactive capacitance is installed.

The total power loss in the uncompensated circuit is the sum of the complex power losses in the transmission line and the load. That is,

$$S_{12} = S_1 + S_2 = I^2 (Z_1 + Z_2).$$

For the new compensated circuit, the sum of all the losses is

$$S_{12N} = S_{1N} + S_2 = I^2 (Z_{1N} + Z_2).$$

Taking their difference,

$$\Delta S = S_{12} - S_{12N} = I^2 (Z_1 + Z_2) - I^2 (Z_{1N} + Z_2).$$

Z_2 cancels out in the terms at the right-hand side. The new result is

$$\Delta S = I^2 (Z_1 - Z_{1N}) = I^2 \left[(R_1 + jX_1) - (R_{1N} - jX_{1N}) \right].$$

The change or saving in complex power does not depend on the load at all. Simplifying further,

$$\Delta S = I^2 (Z_1 - Z_{1N}) = I^2 \left[(R_1 - R_{1N}) + j(X_1 + X_{1N}) \right].$$

The ratio of the savings in power and the old power loss is

$$\frac{\Delta S}{S_{12}} = \frac{I^2 \left[(R_1 - R_{1N}) + j(X_1 + X_{1N}) \right]}{I^2 (R_1 - jX_1)}.$$

Finally,

$$\frac{\Delta S}{S_{12}} = \frac{(R_1 - R_{1N}) + j(X_1 + X_{1N})}{(R_1 - jX_1)}.$$

Notice the addition of the old inductive reactance and the new equivalent capacitive reactance on the second term of the numerator. The addition proves that the savings in power can be significant.

To test the savings in power, four measurements are required. The first two measurements are made before the shunt capacitive reactance is installed. In the measurements, a power meter should be used to measure the complex power at the starting point and ending point of the line. The difference between the two measurements should give the power losses (real and reactive).

The last two measurements and calculations are the same as the first two measurements except that they should be performed after installing the shunt capacitance.

CHAPTER 6

THE SERIES INDUCTANCE OF A

TRANSMISSION LINE

The previous chapters show the importance of inductive reactance in controlling power flow. This chapter shows the derivation of the inductance, voltage drops, and inductive reactance of various configurations of transmission lines.

6.1 Inductance of a Conductor

Consider a cross section of a conductor as shown on Figure 6.1. The total inductance consists of the inductance inside the conductor and outside the conductor. That is,

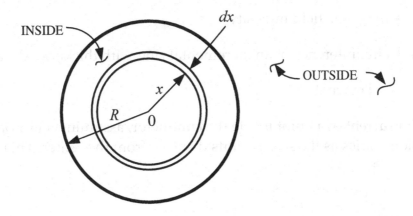

Figure 6.1 Cross Section of a Conductor

$$L = L_i + L_0$$

where

L = total inductance,

L_i = inductance inside the conductor (or inner inductance), and

L_0 = inductance outside the conductor (or outer inductance).

6.1.1 Inner inductance

The line integral of the magnetic field intensity, from $x = 0$ to $x = R$, is the total current

$$\oint_x H \cdot dl = i$$

Where

H = magnetic field intensity,

dl = circumference of an incremental ring with thickness dx, and

i = total current.

The current of a point inside the conductor, assuming a uniform distribution, varies as the square of its distance from the center. That is, the current at x is

$$i_x = \left(\frac{x}{R}\right)^2 i.$$

Since $dl = 2\pi dx$

$$\oint_x H \cdot dl = |H| 2\pi x = i_x .$$

Therefore

$$|H| 2\pi x = \left(\frac{x}{R}\right)^2 i .$$

Solving for $|H|$

$$|H| = \left(\frac{1}{2\pi x}\right)\left(\frac{x}{R}\right)^2 i = \frac{xi}{2\pi R^2} .$$

The magnetic field intensity at a point x distance from the center is the product of the distance and the current divided by the constant $2\pi R^2$. As the distance increases so its does the magnetic field intensity.

The energy, due to the magnetic field, inside one-half of the circle is

$$W_{mf}(half) = \frac{1}{2}\mu\mu_0 \int |H|^2 \, dv .$$

For the whole circle,

$$W_{mf} = \mu\mu_0 \int |H|^2 \, dv = \mu\mu_0 \int_0^R \left(\frac{xi}{2\pi R^2}\right)^2 2\pi x dx = \frac{\mu\mu_0}{8\pi} i^2 .$$

From circuit analysis,

$$W_{mf} = \frac{1}{2} L_i i^2 .$$

Equating the last two equations and solving for the inductance inside the conductor, the inductance inside the conductor is

$$L_i = \frac{\mu\mu_0}{4\pi}.$$

The result shows that the inner inductance is independent of the radius. It is purely a function of the material properties.

In summary, the derivation of the inner inductance consists of four equations. They are:

$$\oint_x H \cdot dl = i \qquad\qquad (1)$$

$$\oint_x H \cdot dl = |H| 2\pi x = i_x \qquad\qquad (1a)$$

$$i_x = \left(\frac{x}{R}\right)^2 i \qquad\qquad (2)$$

$$W_{mf} = \mu\mu_0 \int |H|^2 \, dv = \mu\mu_0 \int_0^R \left(\frac{xi}{2\pi R^2}\right)^2 2\pi x \, dx = \frac{\mu\mu_0}{8\pi} i^2 \qquad (3)$$

$$W_{mf} = \frac{1}{2} L i^2 \qquad\qquad (4)$$

Two of the four equations relate current in the conductor with the magnetic field intensity inside the conductor and the quadratic distribution of current inside the conductor. The other two equations relate the magnetic field energy from the magnetic field intensity and circuit property of inductance.

6.1.2 Outer inductance

The outer inductance in a conductor is produced by neighboring conductor as shown on Figure 6.2.

NOTE: CURRENTS IN THE CONDUCTORS ARE
IN OPPOSITE DIRECTION.

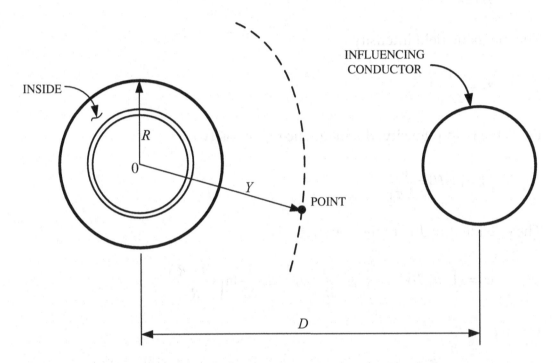

Figure 6.2 Representation of Influencing Conductor for Calculating Outer Inductance

The current due to the outside inductance is

$$\oint H \cdot dl = i$$

where

H = magnetic field intensity due to outside inductance,

dl = incremental ring of radius y, and

i = current.

Performing the line integral,

$$H(2\pi y) = i.$$

The magnetic field intensity is

$$H = \frac{i}{2\pi y}.$$

From the field intensity, the magnetic flux density is

$$|B| = u_0 |H| = \frac{u_0 i}{2\pi y}.$$

The outside flux due to the conductor is

$$\varphi_0 = \int_{R}^{D-R} \mu_0 |H| dy = \int_{R}^{D-R} \mu_0 \frac{i}{2\pi y} dy = \mu_0 \frac{i}{2\pi} \ln\left(\frac{D-R}{R}\right).$$

When $D \gg R$,

$$\varphi_0 = \mu_0 \frac{i}{2\pi} \ln\left(\frac{D}{R}\right).$$

For two conductors, the flux is twice of the single conductor. That is,

$$\varphi_0 = 2\left[\mu_0 \frac{i}{2\pi} \ln\left(\frac{D}{R}\right)\right] = \mu_0 \frac{i}{\pi} \ln\left(\frac{D}{R}\right).$$

The outside flux is also proportional to the current. That is,

$$\varphi_0 = L_0 i \,.$$

Equating the last two equations and solving for L_0,

$$L_0 = \frac{\varphi_0}{i} = \frac{\mu_0}{\pi} \ln\left(\frac{D}{R}\right) \qquad\qquad D \gg R \,, \text{2- conductors}$$

Note that the outer inductance depends on its radius and the distance between the two conductors.

In summary, the derivation of the outer inductance also consists of four independent equations. These equations are shown below:

$$\oint H \cdot dl = i \tag{1}$$

$$H(2\pi y) = i \tag{1a}$$

$$|B| = u_0 |H| = \frac{u_0 i}{2\pi y} \qquad \text{(flux density)} \tag{2}$$

$$\varphi_0 = \int_{R}^{D-R} \mu_0 |H| dy = \int_{R}^{D-R} \mu_0 \frac{i}{2\pi y} dy = \mu_0 \frac{i}{2\pi} \ln\left(\frac{D-R}{R}\right) \quad \text{(total flux)} \tag{3}$$

$$\varphi_0 = \mu_0 \frac{i}{2\pi} \ln\left(\frac{D}{R}\right) \qquad \text{(for 1 conductor)}$$

$$\tag{3a}$$

$$\varphi_0 = 2\left[\mu_0 \frac{i}{2\pi} \ln\left(\frac{D}{R}\right)\right] = \mu_0 \frac{i}{\pi} \ln\left(\frac{D}{R}\right) \quad \text{(for 2 conductors)} \tag{3b}$$

$$\varphi_0 = L_0 i \qquad \text{(proportionality constant)} \tag{4}$$

Like the derivation for inner inductance, the derivation for outer inductance starts with line integral of magnetic field intensity. It is then followed by the magnetic flux density, the total flux, and finally the outer inductance as the proportionality constant between the total flux and current flow.

6.1.3 Total inductance

The total inductance is the sum of the inner inductance and the outside inductance. That is,

$$L = L_i + L_0 = \frac{\mu\mu_0}{4\pi} + \frac{\mu_0}{\pi}\ln\left(\frac{D}{R}\right) \qquad\qquad D \gg R \text{ , 2-conductors}$$

Factoring $\frac{\mu_0}{2\pi}$ and noting that

$$2\ln\left(\frac{D}{R}\right) = \ln\left(\frac{D}{R}\right)^2 = \ln\left(\frac{1}{D}\right)^{-2} + \ln\left(\frac{1}{R}\right)^2$$

the total inductance is

$$L = \frac{\mu_0}{2\pi}\left[\left(\frac{\mu}{4} + \ln\frac{1}{R}\right) + \left(\frac{\mu}{4} + \ln\frac{1}{R}\right) - 2\ln\frac{1}{D}\right].$$

The first two terms inside the square brackets are the self inductances of the two conductors. Its last term is the mutual inductance. That is,

$$L_1 = L_2 \triangleq \frac{\mu_0}{2\pi}\left(\frac{\mu}{4} + \ln\frac{1}{R}\right).$$

Note that self inductance is different from inner inductance.

The mutual inductance is defined as

$$M_{12} \triangleq \frac{\mu_0}{2\pi}\left(\ln\frac{1}{D}\right).$$

Hence, the total inductance is

$$L = L_1 + L_2 - 2M_{12}.$$

Note that the above derivation uses two conductors. The same procedure may be used to derive the total inductance when there are more than two conductors.

6.2 Voltage Drop across a Conductor

The formula for the total inductance may be used in finding the voltage drop across a conductor.

Consider two conductors with opposite current flow as shown on Figure 6.3.

The phasor voltage drop is

$$\Delta V = IZ = j\omega(L)I = j\omega I\left(L_1 + L_2 - 2M_{12}\right).$$

The mutual inductance can be broken into two terms resulting in

$$\Delta V = j\omega I\left(L_1 + L_2 - M_{12} - M_{12}\right) = j\omega\left(L_1 I + L_2 I - M_{12}I - M_{12}I\right).$$

Since

$$I_1 + I_2 = 0$$

then

$$I_1 = -I_2 = I.$$

NOTE: VOLTAGE ACROSS THE
CONDUCTOR CAN BE LINE-TO-LINE, OR
LINE-TO-GROUND (SINGLE PHASE).

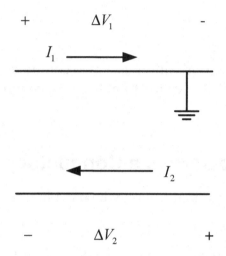

Figure 6.3 The Voltage Drop across a Conductor

Using the current with the same subscript as the inductance,

$$\Delta V = j\omega\left[L_1 I_1 + L_2\left(-I_2\right) - M_{12} I_1 - M_{12}\left(-I_2\right)\right].$$

Re-grouping the terms

$$\Delta V = j\omega\left[\left(L_1 I_1 + M_{12} I_2\right) - \left(L_2 I_2 + M_{12} I_1\right)\right].$$

The terms inside the first parenthesis correspond to the voltage drop across the first conductor. That is,

$$\Delta V_1 = j\omega\left(L_1 I_1 + M_{12} I_2\right).$$

Similarly, the voltage drop across the second conductor is

$$\Delta V_2 = j\omega \left(L_2 I_2 + M_{12} I_1 \right).$$

The above results can be generalized. For n-conductors, the voltage drop across the vth conductor is

$$\Delta V_v = j\omega I_v L_v + j\omega \sum_{\substack{u=1 \\ u \neq v}}^{n} M_{vu} I_u.$$

6.3 Voltage Drop on an Untransposed 3-Phase System

The formula developed in section 6.1.3 system may be generalized as follows:

$$L_v \triangleq \frac{\mu_0}{2\pi} \left(\frac{\mu_v}{4} + \ln \frac{1}{R_v} \right)$$

and

$$M_{vu} \triangleq \frac{\mu_0}{2\pi} \left(\ln \frac{1}{D_{vu}} \right)$$

where

L_v = inductance of the vth conductor,

M_{vu} = mutual inductance between the vth and u^{th} conductors,

R_v = radius of the vth conductor, and

D_{vu} = distance between the vth and u^{th} conductors.

In section 6.2, the voltage drop across a conductor may also be generalized as follows

$$\Delta V_v = j\omega I_v L_v + \sum_{\substack{u=1 \\ u \neq v}}^{n} j\omega I_u M_{vu}.$$

Figure 6.4 shows the schematic representation of the system. Note that the sum of all the currents is zero. That is,

$$I_1 + I_2 + I_3 = 0.$$

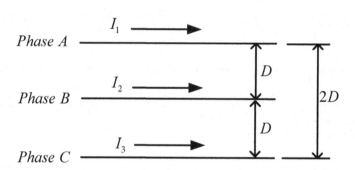

Figure 6.4 An Untransposed Three-Phase System

By carefully following the formula, the voltage drop across a conductor of the system can be found. That is,

$$\Delta V_A = j\omega I_1 \left[\frac{\mu_0}{2\pi}\left(\frac{\mu}{4}+\ln\frac{1}{R}\right)\right] + j\omega I_2\left[\frac{\mu_0}{2\pi}\ln\frac{1}{D}\right] + j\omega I_3\left[\frac{\mu_0}{2\pi}\ln\frac{1}{2D}\right]$$

$$\Delta V_B = j\omega I_2 \left[\frac{\mu_0}{2\pi}\left(\frac{\mu}{4}+\ln\frac{1}{R}\right)\right] + j\omega I_1\left[\frac{\mu_0}{2\pi}\ln\frac{1}{D}\right] + j\omega I_3\left[\frac{\mu_0}{2\pi}\ln\frac{1}{D}\right]$$

$$\Delta V_C = j\omega I_3 \left[\frac{\mu_0}{2\pi}\left(\frac{\mu}{4}+\ln\frac{1}{R}\right)\right] + j\omega I_2 \left[\frac{\mu_0}{2\pi}\ln\frac{1}{D}\right] + j\omega I_1 \left[\frac{\mu_0}{2\pi}\ln\frac{1}{2D}\right].$$

Notice that ΔV_B, unlike the others, can't have $\ln\dfrac{1}{2D}$. Since $I_1 + I_2 + I_3 = 0$, then

$$I_3 = -\left(I_1 + I_2\right).$$

Substituting the above,

$$\Delta V_A = j\omega I_1 \left[\frac{\mu_0}{2\pi}\left(\frac{\mu}{4}+\ln\frac{2D}{R}\right)\right] + j\omega I_2 \left[\frac{\mu_0}{2\pi}\ln 2\right]$$

$$\Delta V_B = j\omega I_2 \left[\frac{\mu_0}{2\pi}\left(\frac{\mu}{4}+\ln\frac{D}{R}\right)\right]$$

$$\Delta V_C = j\omega I_3 \left[\frac{\mu_0}{2\pi}\left(\frac{\mu}{4}+\ln\frac{2D}{R}\right)\right] + j\omega I_2 \left[\frac{\mu_0}{2\pi}\ln 2\right].$$

Phase B has the least voltage drop. Equivalently, phase B has the least inductance.

6.4 Voltage Drop on a Transposed 3-Phase System

Figure 6.5 shows the representation of a transposed system. It has three sections. In each section, the position of a phase differs from the other positions. Since phase occupies the middle position exactly once, it can be inferred that the voltage drops across each conductor are equal. A transposed 3-phase system conserves energy since there is no unbalanced inductance among the phases.

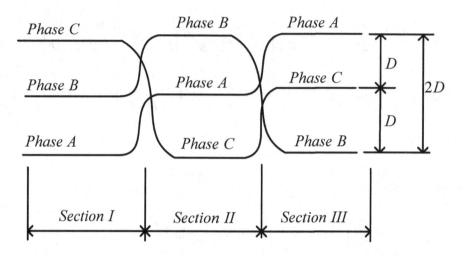

Figure 6.5 A Transposed Three-Phase System

The voltage drops of phase A in each section are shown by

$$\Delta V_A(\text{section I}) = j\omega I_1 \left[\frac{\mu_0}{2\pi}\left(\frac{\mu}{4} + \ln\frac{1}{R} \right) \right] + j\omega I_2 \left[\frac{\mu_0}{2\pi}\ln\frac{1}{D} \right] + j\omega I_3 \left[\frac{\mu_0}{2\pi}\ln\frac{1}{2D} \right]$$

$$\Delta V_A(\text{section II}) = j\omega I_1 \left[\frac{\mu_0}{2\pi}\left(\frac{\mu}{4} + \ln\frac{1}{R} \right) \right] + j\omega I_2 \left[\frac{\mu_0}{2\pi}\ln\frac{1}{D} \right] + j\omega I_3 \left[\frac{\mu_0}{2\pi}\ln\frac{1}{D} \right]$$

$$\Delta V_A(\text{section III}) = j\omega I_1 \left[\frac{\mu_0}{2\pi}\left(\frac{\mu}{4} + \ln\frac{1}{R} \right) \right] + j\omega I_2 \left[\frac{\mu_0}{2\pi}\ln\frac{1}{2D} \right] + j\omega I_3 \left[\frac{\mu_0}{2\pi}\ln\frac{1}{D} \right]$$

Notice that the terms of the sections are identical with each other except for the denominator in $\frac{1}{D}$ or $\frac{1}{2D}$. In section II, the conductor has no $\frac{1}{2D}$ factor.

6.5   Inductive Reactance of a Two-Conductor, Bundled, Transposed 3-Phase System

In a bundled 3-phase system, there are two conductors per phase. See Figure 6.6.

NOTE: EACH CONDUCTOR HAS RADIUS, R.

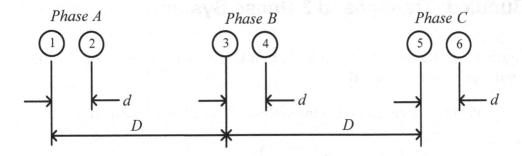

Figure 6.6 A Bundled 3-Phase System

The voltage drop across conductor number 1 is

$$\Delta V_A(\text{conductor 1}) = j\omega \frac{I_A}{2}\left[\frac{\mu_0}{2\pi}\left(\mu + \ln\frac{1}{R}\right)\right] + j\omega\frac{I_A}{2}\left[\frac{\mu_0}{2\pi}\ln\frac{1}{d}\right]$$

$$+ j\omega\frac{I_B}{2}\left[\frac{\mu_0}{2\pi}\ln\frac{1}{D}\right] + j\omega\frac{I_B}{2}\left[\frac{\mu_0}{2\pi}\ln\frac{1}{D+d}\right]$$

$$+ j\omega\frac{I_C}{2}\left[\frac{\mu_0}{2\pi}\ln\frac{1}{2D}\right] + j\omega\frac{I_C}{2}\left[\frac{\mu_0}{2\pi}\ln\frac{1}{2D+d}\right].$$

Again, the terms differ in the current and denominator of the natural logarithmic function.

Without derivation the inductive reactive reactance of a bundled two-conductor 3-phase system with perfect transposition is

$$X_L = \frac{\mu_0 \omega}{2\pi} \left[\frac{1}{8} + \ln\left(\frac{\sqrt[3]{2}}{\sqrt{Rd}} D \right) \right] \text{ ohm per meter.}$$

6.6 Inductive Reactance of a Four-Conductor, Bundled, Transposed 3-Phase System

Figure 6.7 is the case of 4-conductor, bundled 3-phase system. Again, the conductors are transposed.

Without derivation, the inductive reactance of the system is

$$X_L = \frac{\mu_0 \omega}{2\pi} \left[\frac{1}{16} + \ln\left(\frac{\sqrt[3]{2}}{\sqrt[4]{Rd_{12}d_{13}d_{14}}} D \right) \right].$$

Notice the similarity of the formula with the 2-conductor, bundled, transposed case. The inductive reactance of the 4-conductor bundled, transposed case is smaller than that of the 2-conductor, bundled, transposed case (by approximately one-half).

NOTE: EACH CONDUCTOR HAS RADIUS, R.

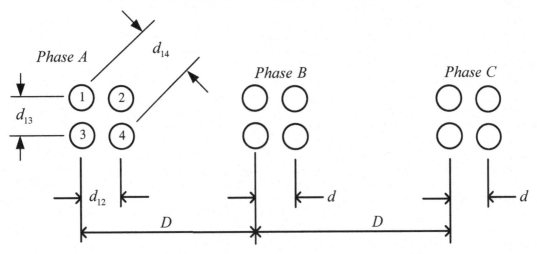

Figure 6.7 A Four-Conductor Bundled Transposed System

CHAPTER 7

SHUNT CAPACITANCE OF A

TRANSMISSION LINE

While inductance is related to the flow of current in a conductor, capacitance is related to the difference in the voltages of two charges when a point is at the edge of a charge. A transmission line represents a charge. There is therefore a capacitance between two transmission lines; or, between a transmission line and the earth (since the earth is also a charge). The method of images will be used in calculating the capacitance not only due to the presence of a neighboring wire but of the earth as well.

7.1 Capacitance between Two Charges (or Conductors)

Figure 7.1 shows two oppositely-charged charges separated D. A charge can be a current-carrying conductor. The $+Q$ charge is at distance R_1 from the given point. Similarly, the $-Q$ charge is separated by the distance R_2 from the same point.

The voltage at the point is given by

$$V_P = \frac{+Q}{4\pi\varepsilon_0}\ln\left(\frac{1}{R_1}\right)^2 + \frac{-Q}{4\pi\varepsilon_0}\ln\left(\frac{1}{R_2}\right)^2.$$

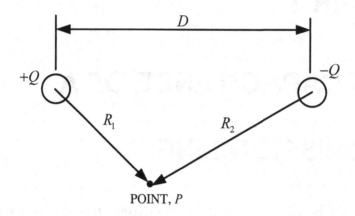

Figure 7.1 A Point Between Two Opposite Charges

After simplifying the equation, the voltage at the point is

$$V_P = \frac{Q}{2\pi\varepsilon_0} \ln\frac{R_2}{R_1}.$$

Suppose the point is moved closed to the positive charge as shown on Figure 7.2.

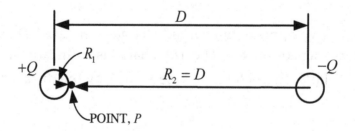

Figure 7.2 Condition when the Point is at the Edge of $+Q$

Now, the voltage at the point is

$$V_1 = \frac{Q}{2\pi\varepsilon_0} \ln \frac{D}{R_1}.$$

Moving the point close to $-Q$ (see Figure 7.3), the voltage is

$$V_2 = \frac{Q}{2\pi\varepsilon_0} \ln \frac{R_2}{D}.$$

The difference between the two voltages is

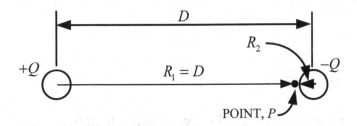

Figure 7.3 Case when the Point is at the Edge of $-Q$

$$V_1 - V_2 = \frac{Q}{2\pi\varepsilon_0} \ln \frac{D}{R_1} - \frac{Q}{2\pi\varepsilon_0} \ln \frac{R_2}{D} = \frac{Q}{2\pi\varepsilon_0} \left(\ln \frac{D}{R_1} - \ln \frac{R_2}{D} \right).$$

Since

$$-\ln \frac{R_2}{D} = +\ln \frac{D}{R_2},$$

$$V_1 - V_2 = \frac{Q}{2\pi\varepsilon_0} \left(\ln \frac{D}{R_1} + \ln \frac{D}{R_2} \right) = \frac{Q}{2\pi\varepsilon_0} \left(\ln \frac{D^2}{R_1 R_2} \right).$$

Simplifying further,

$$V_1 - V_2 = \frac{Q}{2\pi\varepsilon_0} \ln\left(\frac{D}{\sqrt{R_1 R_2}}\right)^2.$$

Finally,

$$V_1 - V_2 = \frac{Q}{\pi\varepsilon_0} \ln\left(\frac{D}{\sqrt{R_1 R_2}}\right).$$

When radii of the two charges are equal, $R_1 = R_2 = R$, the point experiences the voltage

$$V_1 - V_2 = \frac{Q}{\pi\varepsilon_0} \ln\left(\frac{D}{R}\right).$$

The capacitance between the two charges is defined as

$$C \cong \frac{Q}{V}.$$

Substituting $V = V_1 - V_2$

$$C = \frac{Q}{\dfrac{Q}{\pi\varepsilon_0} \ln\left(\dfrac{D}{R}\right)} = \frac{\pi\varepsilon_0}{\ln\left(\dfrac{D}{R}\right)} \qquad \text{(condition: } R_1 = R_2 = R\text{)}$$

The capacitance between the two oppositely charges depends on their separation. Furthermore, it is assumed that the capacitance is measured exactly between the charges.

Review again the voltage at a point from two charges. If there are n charges, its voltage is

100

$$V_P = \sum_{v=1}^{n} \frac{Q}{2\pi\varepsilon_0} \ln\frac{1}{R_v}$$

where the sign of Q determines the sign of a term in the summation. A general equation may also be derived for the capacitance.

7.2 The Method of Images in Determining Capacitance

Figure 7.4 shows the configuration of the charges using the method of images. The method includes the effect of the earth in calculating the capacitance between two charges or conductors. On the figure, charges of the same sign are diagonally opposite of each other.

From the previous section, the voltage at the point is

$$V_P = \sum_{v=1}^{n} \frac{Q}{2\pi\varepsilon_0} \ln\frac{1}{R_v} .$$

Expanding the formula,

$$V_P = \frac{Q}{2\pi\varepsilon_0} \ln\frac{1}{R_1} + \frac{-Q}{2\pi\varepsilon_0} \ln\frac{1}{R_2} + \frac{-Q}{2\pi\varepsilon_0} \ln\frac{1}{R_3} + \frac{Q}{2\pi\varepsilon_0} \ln\frac{1}{R_4} .$$

When the point is at the edge of the first $+Q$ (see Figure 7.5),

$$V_1 = \frac{Q}{2\pi\varepsilon_0} \ln\frac{1}{R_1} + \frac{-Q}{2\pi\varepsilon_0} \ln\frac{1}{D} + \frac{-Q}{2\pi\varepsilon_0} \ln\frac{1}{2H} + \frac{Q}{2\pi\varepsilon_0} \ln\frac{1}{\sqrt{D^2 + (2H)^2}} .$$

Doing the same for the point at the edge of the second charge, $-Q$,

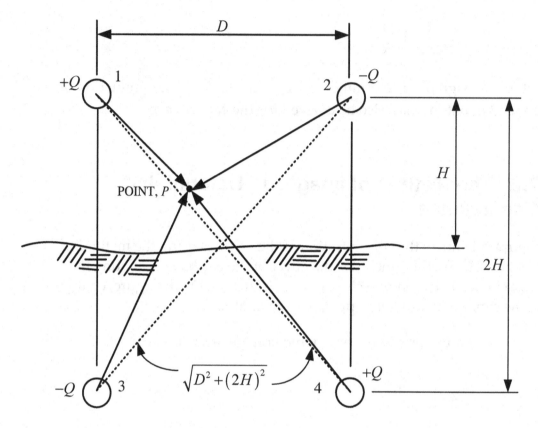

Figure 7.4 Configuration of the Charges in the Method of Images

$$V_2 = \frac{Q}{2\pi\varepsilon_0} \ln\frac{1}{D} + \frac{-Q}{2\pi\varepsilon_0} \ln\frac{1}{R_2} + \frac{Q}{2\pi\varepsilon_0} \ln\frac{1}{2H} + \frac{-Q}{2\pi\varepsilon_0} \ln\frac{1}{\sqrt{D^2 + (2H)^2}}.$$

With the radii of the charges equal, i.e. $R_1 = R_2 = R$,

$$V_1 - V_2 = \frac{Q}{2\pi\varepsilon_0} \left(2\ln\frac{1}{R} - 2\ln\frac{1}{D} - 2\ln\frac{1}{2H} + 2\ln\frac{1}{\sqrt{D^2 + (2H)^2}} \right).$$

Simplifying further,

$$V_1 - V_2 = \frac{Q}{\pi \varepsilon_0} \left(\ln \frac{1}{R} + \ln D + \ln (2H) + \ln \frac{1}{\sqrt{D^2 + (2H)^2}} \right).$$

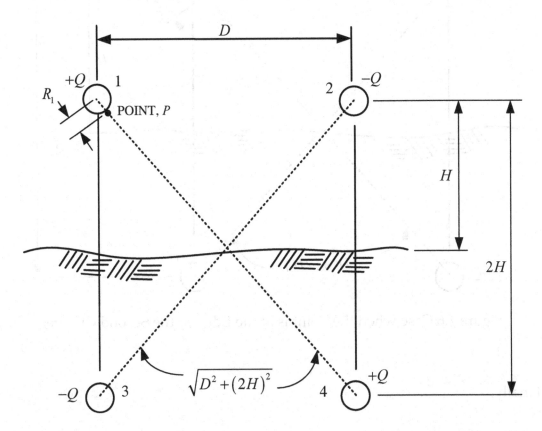

Figure 7.5 Case when the Point is at the Edge of the First Charge

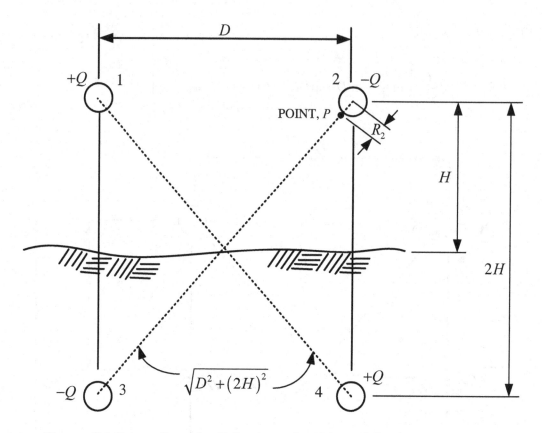

Figure 7.6 Case when the Point is at the Edge of the Second Charge

Finally,

$$V_1 - V_2 = \frac{Q}{\pi\varepsilon_0}\ln\left(\frac{D}{R}\frac{2H}{\sqrt{D^2+(2H)^2}}\right).$$

Using $C \cong \dfrac{Q}{V}$,

$$C = \frac{\pi \varepsilon_0}{\ln\left(\dfrac{D}{R}\dfrac{2H}{\sqrt{D^2+(2H)^2}}\right)} \quad \text{Farad/meter}$$

Notice that the method of images differs from the normal method by the factor of $\dfrac{2H}{\sqrt{D^2+(2H)^2}}$ in the logarithmic function at the denominator.

The above capacitance is the capacitance between two charges (or transmission lines) with the effect of the earth. Notice how the factor $\dfrac{2H}{\sqrt{D^2+(2H)^2}}$ makes the factor $\dfrac{D}{R}$ smaller. As a result, the earth increases the capacitance of the line.

7.3 Capacitance of a Three-Phase System

The formula $V_P = \sum\limits_{v=1}^{n} \dfrac{Q}{2\pi\varepsilon_0} \ln\dfrac{1}{R_v}$ will be used again in finding the capacitance of a three-phase system. See Figure 7.7

For the voltage at the first charge,

$$V_1 = \sum_{v=1}^{6} \frac{Q}{2\pi\varepsilon_0} \ln\left(\frac{1}{R_v}\right)$$

or

$$V_1 = \frac{Q_1}{2\pi\varepsilon_0} \ln\frac{1}{R} + \left(\frac{-Q_1}{2\pi\varepsilon_0} \ln\left(\frac{1}{2H}\right)\right)$$

105

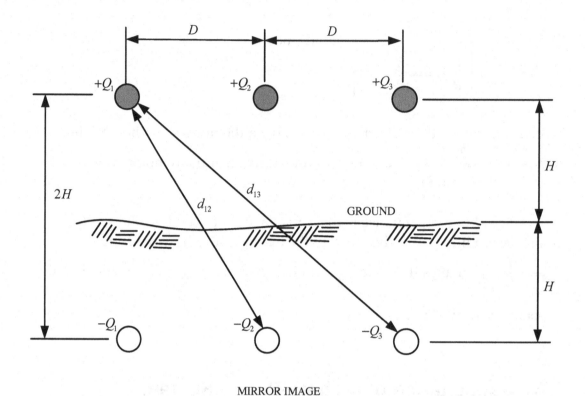

Figure 7.7 Configuration of Charges for Finding the Capacitance of a 3-Phase System

$$+\frac{Q_2}{2\pi\varepsilon_0}\ln\frac{1}{D}+\left(\frac{-Q_2}{2\pi\varepsilon_0}\ln\frac{1}{d_{12}}\right)$$

$$+\frac{Q_3}{2\pi\varepsilon_0}\ln\frac{1}{2D}+\left(\frac{-Q_3}{2\pi\varepsilon_0}\ln\left(\frac{1}{d_{13}}\right)\right)$$

Since

$$\frac{Q_1}{2\pi\varepsilon_0}\ln\left(\frac{1}{R_1}\right)+\left(\frac{-Q_1}{2\pi\varepsilon_0}\ln\left(\frac{1}{2H}\right)\right)=\frac{Q_1}{2\pi\varepsilon_0}\ln\left(\frac{2H}{R_1}\right)$$

then

$$V_1=\frac{Q_1}{2\pi\varepsilon_0}\ln\left(\frac{2H}{R_1}\right)+\frac{Q_2}{2\pi\varepsilon_0}\ln\left(\frac{d_{12}}{D}\right)+\frac{Q_3}{2\pi\varepsilon_0}\ln\left(\frac{d_{13}}{2D}\right).$$

Similarly,

$$V_2=\frac{Q_1}{2\pi\varepsilon_0}\ln\left(\frac{d_{21}}{D}\right)+\frac{Q_2}{2\pi\varepsilon_0}\ln\left(\frac{2H}{R}\right)+\frac{Q_3}{2\pi\varepsilon_0}\ln\left(\frac{d_{23}}{2D}\right),$$

and

$$V_3=\frac{Q_1}{2\pi\varepsilon_0}\ln\left(\frac{d_{31}}{2D}\right)+\frac{Q_2}{2\pi\varepsilon_0}\ln\left(\frac{d_{32}}{D}\right)+\frac{Q_3}{2\pi\varepsilon_0}\ln\left(\frac{2H}{R}\right).$$

If $Q_1=Q_2=Q_3=Q$ are equal, then the capacitance can be directly calculated from

$$C=\frac{Q}{V}.$$

That is, Q will cancel out in the numerator and the denominator.

CHAPTER 8

MODELS OF HIGH-VOLTAGE

TRANSMISSION LINES

The previous chapter shows how to calculate the inductance and capacitance of power transmission lines. They are required in the models of the lines. Three models are shown here. They are the short line, medium length, and the long-line models.

8.1   Review of Some Mathematics

This chapter uses matrices to represent a system of simultaneous equations. The equations are for voltage and current. To achieve compactness and generality, models of transmission lines are represented in matrix equations.

The long-line model of transmission lines uses second order differential equation. It is, perhaps, the most complicated of all the models. It is, however, the most accurate. A review of derivatives and differential equation is fitting here.

8.1.1   Matrices

A system of two equations may be represented by

$$y_1 = a_1 x_1 + b_1 x_2$$

$$y_1 = a_2 x_2 + b_2 x_2 .$$

In matrix form, the same system is

$$\begin{bmatrix} y_1 \\ y_2 \end{bmatrix} = \begin{bmatrix} a_1 & b_1 \\ a_2 & b_2 \end{bmatrix} \begin{bmatrix} x_1 \\ x_2 \end{bmatrix}.$$

The matrix on the left is one-dimensional. One-dimensional matrix is called a vector. There are three matrices above. They are the constant vector $\begin{bmatrix} y_1 \\ y_2 \end{bmatrix}$, the coefficient matrix $\begin{bmatrix} a_1 & b_1 \\ a_2 & b_2 \end{bmatrix}$, and the unknown vector $\begin{bmatrix} x_1 \\ x_2 \end{bmatrix}$. The unknown vector is always the last matrix on the right-hand side.

In this chapter, the equations for the sending end voltage and current, of a transmission line, will be represented be represented in matrix form. The coefficient matrix are the parameters of the transmission line, and the unknown vector is the receiving end voltage and current.

Not only the matrix representation useful in finding the solution of a system of equations. They are also useful in finding coefficients of a single equation.

Suppose a line is represented by

$$y = a + bx.$$

In this case, the unknowns are the coefficients a and b. Two points are required to find the particular equation. In matrix form,

$$\begin{bmatrix} y_1 \\ y_2 \end{bmatrix} = \begin{bmatrix} 1 & x_1 \\ 1 & x_2 \end{bmatrix} \begin{bmatrix} a \\ b \end{bmatrix}$$

where (x_1, y_1) and (x_2, y_2) are the known (measured) coordinates of the points. The unknown vector $\begin{bmatrix} a \\ b \end{bmatrix}$ can be found by solving the matrix equation.

As an example, suppose the first point has $x_1 = 0$ and $y_1 = 1$, and the second point has $x_2 = 1$ and $y_2 = 3$, then the particular equation gives $a = 1$ and $b = 2$. Therefore, the particular equation of the line is $y = 1 + 2x$.

The above procedure can be used not only on equations of a line but also on non-linear equations (such as the magnetization curve) provided terms or factors that are non-linear can be approximated by their series expansion.

As will be seen, it is possible to calculate the unknown parameters of a transmission line if the measurements at the sending end and receiving end are known.

8.1.2　Derivatives and Differential Equations

Suppose y is a function of x. The derivative of y relative to x is defined as

$$\frac{dy}{dx} = \lim_{\Delta x \to 0} \frac{\Delta y}{\Delta x} = \lim_{\Delta x \to 0} \frac{f(x + \Delta x) - f(x)}{\Delta x}.$$

The limiting case, $\Delta x \to 0$ implies that $\frac{dy}{dx}$ is the slope at the point (x, y). Figure 8.1 shows the behavior of the limiting case when the Δx changes from a large to a smaller value.

In general, the derivative is also a function of x. For example, if

$$y = x^2$$

then

$$\frac{dy}{dx} = 2x.$$

It means that at any point (x, y), the derivative of y is twice the x-coordinate.

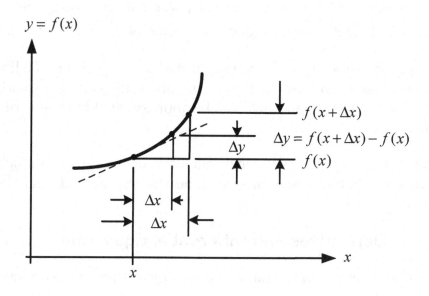

Figure 8.1 The Basic Concept of Derivative

The graph of $\dfrac{dy}{dx} = 2x$ may also be similarly graphed as shown on Figure 8.2. In this case, the graph is a straight line.

The second derivative of y with respect to x is defined as

$$\frac{d^2 y}{dx^2} = \lim_{\Delta x \to 0} \frac{\Delta y'(x)}{\Delta x} = \lim_{\Delta x \to 0} \frac{f'(x + \Delta x) - f'(x)}{\Delta x}.$$

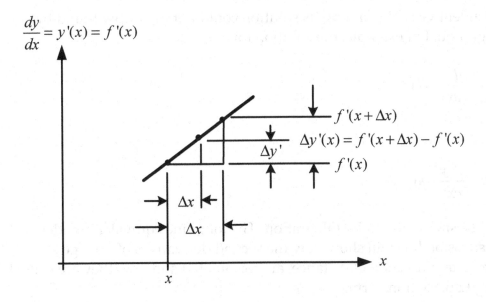

$$\frac{dy}{dx} = y'(x) = f'(x)$$

$f'(x+\Delta x)$

$\Delta y'(x) = f'(x+\Delta x) - f'(x)$

$\Delta y'$

$f'(x)$

Δx

Δx

x

x

Figure 8.2 Graph of the Derivative of $\dfrac{dy}{dx}$ as a Function of x

The most fundamental difference between the first derivative and the second derivative is the use of Δx. While the values of $y's$ (for calculating the first derivative), and $\Delta y's$ (for calculating the second derivative) are different, both must use the same Δx. That is why, in modeling, there is one and only one incremental sample of Δx.

For the example $y = x^2$, its second derivative is

$$\frac{d^2 y}{dx^2} = 2.$$

It shows that the second derivative is a constant. When the second derivative is a constant, the problem of finding the solution becomes a simple matter of integration. However, when the second derivative is a function of the dependent variable, finding its solution could not be simply found by integration. For example, the solution for

$$\frac{d^2y}{dx^2} = ky$$

or

$$\frac{d^2y}{dx^2} - ky = 0$$

can't be simply found by integration. The long-line equivalent model of transmission line will show how the second derivatives of voltage and current, as a function of distance, are formulated and how they are solved using Laplace transform.

8.2   Short Line Equivalent Model

Figure 8.3 shows the equivalent short-line model of a power transmission line. Notice that the model has no shunt capacitive reactance. The model is good for lengths of up to 30 miles.

Using Kirchoff's voltage law,

$$E_S = AE_R + BI_R = (1)E_R + ZI_R.$$

Solving for the current,

$$I_S = CE_R + DI_R.$$

Since there is no element across E_R, the coefficient, C of E_R is zero. That is, $C = 0$. Therefore,

$$C = 0.$$

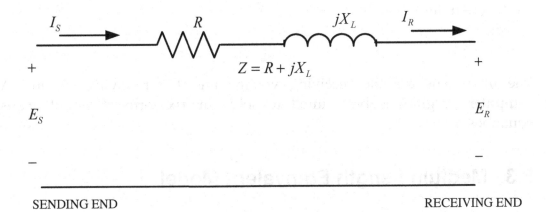

SENDING END   RECEIVING END

Figure 8.3 Short Line Equivalent Model of a High-Voltage Transmission Line

The sending current is equal to the receiving current. Therefore,

$$D = 1.$$

Collecting the voltage and current equations,

$$E_S = AE_R + BI_R$$

$$I_S = CE_R + DI_R.$$

The equations can be formed in matrix form:

$$\begin{bmatrix} E_S \\ I_S \end{bmatrix} = \begin{bmatrix} A & B \\ C & D \end{bmatrix} \begin{bmatrix} E_R \\ I_R \end{bmatrix}$$

where

$$A = 1,$$

$$B = Z.$$

$C = 0$, and

$D = 1$.

The unknowns are the receiving voltage and the receiving current. A computer program is best suited to solve matrix form of simultaneous equations.

8.3 Medium Length Equivalent Model

The medium length equivalent model has shunt capacitance at the sending end and the receiving end. See Figure 8.4. It is good for distances of 30 miles to about 200 miles. In the model, the shunt capacitive reactance is represented by its admittance. The admittance is one-half of the total admittance.

The current in the receiving end is the sum of the two currents. They are the currents in the shunt capacitance and the current in the series impedance of the transmission line.

Using Kirchoff's voltage law,

$$E_S = \left[E_R \left(\frac{Y}{2} \right) \right] Z + I_R Z + E_R$$

where the factor in the first term is the current thru the last shunt capacitance. Simplifying the equation,

$$E_S = \left[\left(1 + \frac{YZ}{2} \right) \right] E_R + ZI_R .$$

Finally,

$$E_S = AE_R + BI_R$$

where

$$A = \left(1 + \frac{YZ}{2}\right), \text{ and}$$

$$B = Z .$$

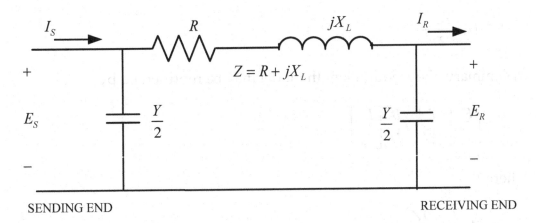

Figure 8.4 The Medium Length Model of a High-Voltage Transmission Line

For the current,

$$I_S = E_S \frac{Y}{2} + E_R \frac{Y}{2} + I_R .$$

Substituting $E_S = \left[\left(1 + \frac{YZ}{2}\right)\right] E_R + ZI_R$ and simplifying gives

$$I_S = \left(\frac{Y^2 Z}{4} + Y\right) E_R + \left(\frac{YZ}{2} + 1\right) I_R .$$

More compactly,

$$I_S = (C)E_R + (D)I_R$$

where

$$C = \frac{Y^2 Z}{4} + Y, \text{ and}$$

$$D = \frac{YZ}{2} + 1.$$

In summary, the medium length model may be represented by

$$\begin{bmatrix} E_S \\ I_S \end{bmatrix} = \begin{bmatrix} A & B \\ C & D \end{bmatrix} \begin{bmatrix} E_R \\ I_R \end{bmatrix}$$

where

$$A = \left(1 + \frac{YZ}{2}\right),$$

$$B = Z,$$

$$C = \frac{Y^2 Z}{4} + Y, \text{ and}$$

$$D = \frac{YZ}{2} + 1.$$

Again, the A, B, C, and D parameters are known. The problem is to find the receiving end voltage and current.

Compare the short line and medium line equivalent models. In the short line model, $C = 0$. The medium line model has $C = \dfrac{Y^2 Z}{4} + Y$. It shows that the medium line model requires more current. Therefore, it has more power loss.

8.4 Long-Line Equivalent Model

The long-line equivalent model uses differential distance, dx of a transmission line. Figure 8.5 is its representation. Note that the origin or $x = 0$ is at the receiving end. Each dx consists of the shunt admittance and the series impedance. The admittance and the impedance are in per unit length (such as ohm per meter). Since the model uses the differential distance from the receiving end, the subscript indicating sending end is now dropped off. Subscript for the receiving end, however, is still used since the end becomes the reference.

The differential voltage across the differential distance is

$$\frac{dE}{dx} = IZ .$$

Part of the current, I flows in the shunt admittance. The differential current in the differential distance is

$$\frac{dI}{dx} = EY .$$

Since the next differential distance will have voltage and current that differs from the first (differential distance), taking their derivatives again gives

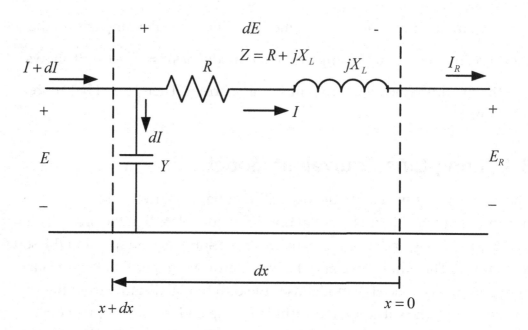

Figure 8.5 The Long-Line Equivalent Model of High-Voltage
Transmission Line

$$\frac{d^2E}{dx^2} = Z\frac{dI}{dx}$$

$$\frac{d^2I}{dx^2} = Y\frac{dE}{dx}.$$

Substituting $\dfrac{dI}{dx} = EY$ into $\dfrac{d^2E}{dx^2} = Z\dfrac{dI}{dx}$ gives

$$\frac{d^2E}{dx^2} - ZYE = 0.$$

The differential voltage across the second differential distance is directly proportional to its admittance, impedance, and the voltage across its terminals.

Taking the Laplace transform of $\dfrac{d^2E}{dx^2} - ZYE = 0$ gives

$$s^2E(s) - sE(x_0) - E'(x_0) - ZYE(s) = 0.$$

At $x = 0$ (initial condition),

$$E(x_0) = E_R,$$

$$I = I_R,$$

and

$$E'(x_0) = \dfrac{dE}{dx}\bigg|_{x=0} = IZ\big|_{x=0} = I_RZ.$$

Substituting the initial conditions in $s^2E(s) - sE(x_0) - E'(x_0) - ZYE(s) = 0$ gives

$$s^2E(s) - sE_R - I_RZ - ZYE(s) = 0.$$

Solving for $E(s)$,

$$E(s) = \dfrac{sE_R}{s^2 - ZY} + \dfrac{I_RZ}{s^2 - ZY}.$$

As a function of x, the solution of $E(s) = \dfrac{sE_R}{s^2 - ZY} + \dfrac{I_RZ}{s^2 - ZY}$ consists of two hyperbolic functions (with the same arguments):

$$E(x) = \left[\cosh\left(x\sqrt{ZY}\right)\right]E_R + \sqrt{\frac{Z}{Y}}\left[\sinh\left(x\sqrt{ZY}\right)\right]I_R .$$

Using the same procedure, the current, as a function of x is

$$I(x) = \sqrt{\frac{Y}{Z}}\left[\sinh\left(x\sqrt{ZY}\right)\right]E_R + \left[\cosh\left(x\sqrt{ZY}\right)\right]I_R .$$

Define the factor, $\sqrt{ZY}$ in the argument of the hyperbolic function, as

$$\sqrt{ZY} = \gamma = \alpha + j\beta .$$

It is called propagation constant and a complex number. Its unit is per meter or m^{-1}.

Similarly, define the constant $\sqrt{\dfrac{Y}{Z}}$ as

$$\sqrt{\frac{Y}{Z}} = Y_0 = \frac{1}{Z_0} .$$

Its inverse, or

$$Z_0 = \sqrt{\frac{Z}{Y}}$$

is called the characteristic impedance of the transmission line. Its dimension is in ohm. Note that the characteristic impedance is the impedance across the differential distance (see Figure 8.5).

Using the constants, the receiving end voltage and current are:

$$E(x) = \left[\cosh\left(\gamma x\right)\right]E_R + Z_0\left[\sinh\left(\gamma x\right)\right]I_R, \quad \text{(from receiving end)}$$

and

$$I(x) = Y_0\left[\sinh\left(\gamma x\right)\right]E_R + \left[\cosh\left(\gamma x\right)\right]I_R. \quad \text{(from receiving end)}$$

The above solutions uses the receiving end as the origin of the distance x. In matrix form, the above equations can be represented as

$$\begin{bmatrix} E(x) \\ I(x) \end{bmatrix} = \begin{bmatrix} A & B \\ C & D \end{bmatrix}\begin{bmatrix} E_R \\ I_R \end{bmatrix} \qquad \text{(from receiving end)}$$

where,

$$A = \cosh\left(\gamma x\right),$$

$$B = Z_0\left[\sinh\left(\gamma x\right)\right],$$

$$C = Y_0\left[\sinh\left(\gamma x\right)\right], \text{ and}$$

$$D = \left[\cosh\left(\gamma x\right)\right].$$

The voltage and current, when the sending end is the origin, may be found by replacing the E_R by E_0, the current I_R by $I(0)$ and subtracting the terms with $\sinh(\ )$. E_0 and $I(0)$ are the (initial) voltage and current. See Figure 8.6.

The solutions for the voltage and current are:

$$E(x) = \left[\cosh\left(\gamma x\right)\right]E_0 - Z_0\left[\sinh\left(\gamma x\right)\right]I(0) \qquad \text{(from sending end)}$$

and

$$I(x) = -Y_0\left[\sinh\left(\gamma x\right)\right]E_0 + \left[\cosh\left(\gamma x\right)\right]I(0). \qquad \text{(from sending end)}$$

123

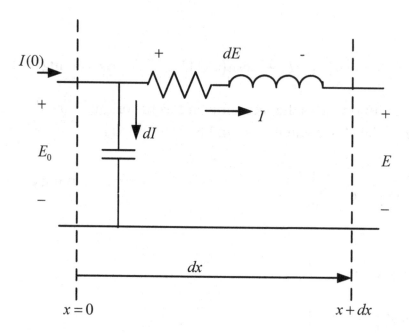

Figure 8.6 Representation of Voltage and Current when the Origin is the Sending End

Similar to receiving end, the above equations can be represented in the following matrix form:

$$\begin{bmatrix} E(x) \\ I(x) \end{bmatrix} = \begin{bmatrix} A & B \\ C & D \end{bmatrix} \begin{bmatrix} E_0 \\ I(0) \end{bmatrix}$$
(from sending end)

where,

$$A = \left[\cosh(\gamma x) \right],$$

$$B = -Z_0 \left[\sinh(\gamma x) \right],$$

$$C = -Y_0 \left[\sinh(\gamma x) \right], \text{ and}$$

$$D = \left[\cosh(\gamma x) \right].$$

8.5 A Test Approach to Measure the Impedance and Shunt Admittance of a Transmission Line

In utilities, the as-built drawings of the transmission lines are sometimes not up to date. The problem is how to find out their impedance and shunt admittance. They may be found using the recommended test approach as described below.

Examine, for example, the equation for the sending voltage of a transmission line using the medium length model. It is repeated here as

$$E_S = \left[\left(1 + \frac{YZ}{2} \right) \right] E_R + ZI_R .$$

The unknowns are $\left[\left(1 + \dfrac{YZ}{2} \right) \right]$ and Z .

The equation has two unknowns. Therefore, two sets of measurements are required. The first set measures the sending voltage, E_S and the receiving voltage-current pair, E_R and I_R . They may be measured at a given load. In the second set, the same variables are measured except at a different load.

Next, substitute the first set of measurements to the equation. Repeat with the second measurements. Two simultaneous equations should result.

From the equations solve for $\left[\left(1+\dfrac{YZ}{2}\right)\right]$ and Z. Once Z is found, Y can be found from the calculated value of $\left[\left(1+\dfrac{YZ}{2}\right)\right]$.

CHAPTER 9

ITERATIVE APPROACH IN SOLVING

TRANSMISSION LINE PROBLEMS

Broadly speaking, a power system consists of generators or power plants, consumer loads or demands, and power transmission networks. Among the three, the latter is the most complicated since their solutions require iterative approach. An electrical engineer, for example, may spend a year performing calculations and analysis of a ten-bus transmission network. Using software with an iterative approach, the same calculations could take only an hour or so. This chapter examines the matrix representation of transmission lines and their solutions using an iterative approach. In particular, the approach uses the Gauss iterative technique in solving simultaneous complex equations.

9.1   Review of Some Mathematics – The Iterative Approach

Iteration is a cycle of calculations. Finding a solution using the iterative approach takes several iterations. In each cycle, the variables take on different values depending on their sequence. A sequence is indicated by an index in the subscript and/or superscript. Once all the values are defined, the variables are combined in the form of a series, which is a sum of the variables.

In analyzing iterative approach, start with the first iteration. Assign values to the indices and determine how variables differ with each other. Next, take the second iteration. Again, assign values to the indices and determine how the variables behave. Oftentimes, the behavior of a variable in the second iteration differs with that of the first iteration. The key is on finding that difference.

As an example, the equation

$$y_i^{(v+1)} = \frac{1}{\left(y_i^{(v)}\right)} - \sum_{\substack{k=1 \\ k \neq i}}^{n} y_k^{(v)}, \ i = 2,3,...,n \text{ and } k = 1,2,3,...,n$$

requires that i starts with 2, and k starts with 1. Since i ranges from 2 to n, then there are infinite number of (variables) y. The subscripts at summation sign show that k can't be equal to i. The superscript $(v+1)$ indicates the iteration number.

The following two equations are samples of the equations for the first iteration:

$$y_2^{(1)} = \frac{1}{\left(y_2^{(0)}\right)} - \left(y_1^{(0)} + y_3^{(0)} + y_4^{(0)} + ...\right)$$

$$y_3^{(1)} = \frac{1}{\left(y_3^{(0)}\right)} - \left(y_1^{(0)} + y_2^{(0)} + y_4^{(0)} + ...\right)$$

Notice the structure of the variables in the samples. The denominator in the first term on the right-hand side can't appear as a term inside the parenthesis.

Samples for the second iteration are:

$$y_2^{(2)} = \frac{1}{\left(y_2^{(1)}\right)} - \left(y_1^{(1)} + y_3^{(1)} + y_4^{(1)} + ...\right)$$

$$y_3^{(2)} = \frac{1}{\left(y_3^{(1)}\right)} - \left(y_1^{(1)} + y_2^{(1)} + y_4^{(1)} + ...\right)$$

The same conclusion as from the first iteration applies. Additionally, values of the variables now take the values from the first iteration. Once the samples for the first iteration are made, it becomes a simple matter of changing the indices in the samples of the second iteration. The process repeats.

9.2 Classification of High-Voltage Transmission Line Busses

There are types of busses. They are the load bus, the voltage control bus, and the reference (also called swing or slack) bus. In all the busses, the complex power demand is known in a bus.

A load bus has the generators and the load. The amount of generated real power and reactive power must be specified in the bus. However, the voltage magnitude and power angle are calculated.

As the name implies, the voltage controller bus controls the voltage. Its voltage magnitude and generated real power must be fixed. The generated reactive power and power angle, however, must be calculated.

The reference (slack or swing) bus is usually the most expensive bus. Its voltage magnitude and power angle must be fixed (e,g, $1\angle0^0$ per unit volt). The amount of generated real power and reactive power must be calculated. In calculations, it is the swing bus that is the most critical bus.

9.3 Matrix Representation of a Generic Two-Source, Three-Bus Transmission System

Figure 9.1 shows a generic transmission system with two sources and three busses. The bus with no generator has reactive power compensation ($jQ3$). A

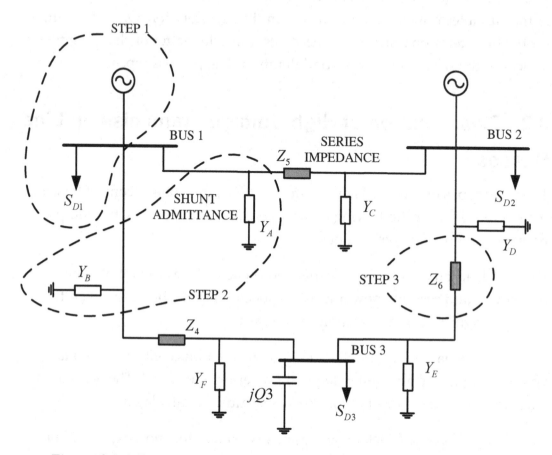

Figure 9.1 A Power System with Two Generators and Three Busses

solid rectangle represents the series impedance of a line. Conversely, a blank rectangle represents shunt admittance (to the ground).

Three steps are required in simplifying the system. They are shown on the figure. They are:

1. Combine the generator and the demand and represent them as bus power,

2. Add shunt admittances that are in parallel with each other, and

3. Replace all series impedances by their admittances.

The bus powers, indicated by one-input AND gate, are represented by S_1, S_2, and S_3. All shunt admittances are the sum of shunt admittances that are in parallel with each other. That is,

$$Y_1 = Y_A + Y_B,$$

$$Y_2 = Y_C + Y_D, \text{ and}$$

$$Y_3 = Y_E + Y_F.$$

All series impedances are converted to their admittances. That is,

$$Y_4 = \frac{1}{Z_4},$$

$$Y_5 = \frac{1}{Z_5}, \text{ and}$$

$$Y_6 = \frac{1}{Z_6}.$$

Figure 9.2 incorporates all of the above simplifications.

The current in a node is zero. For the first bus power,

$$I_1 = V_1 Y_1 + (V_1 - V_2)Y_5 + (V_1 - V_3)Y_4.$$

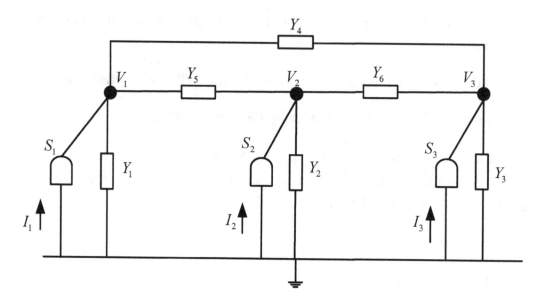

Figure 9.2 Simplified Diagram of Figure 9.1

Similarly, for the second and third sources,

$$I_2 = V_2 Y_2 + (V_2 - V_1)Y_5 + (V_2 - V_5)Y_6,$$

and

$$I_3 = V_3 Y_3 + (V_3 - V_1)Y_4 + (V_3 - V_2)Y_6.$$

Expanding the right-hand side of each current equation and simplifying according to increasing subscripts of the voltages give

$$I_1 = (Y_1 + Y_4 + Y_5)V_1 - Y_5 V_2 - Y_4 V_3,$$

$$I_2 = -Y_5 V_1 + (Y_2 + Y_5 + Y_6)V_2 - Y_6 V_3,$$

and

$$I_3 = -Y_4V_1 - Y_6V_2 + (Y_3 + Y_4 + Y_6)V_3$$

The patterns of the equations are as follows:

- There are exactly two terms with negative sign in each equation, and

- There is exactly one term with one positive sign.

Admittances of terms with positive sign, such as $(Y_1 + Y_4 + Y_5)$, are called driving point admittances. They are all connected to the same bus.

Admittance from node i to node j can be represented as y_{ij}. Similarly, all driving point admittances may represented as y_{ii}. With these conventions, the above equations can be represented as

$$I_1 = y_{11}V_1 + y_{12}V_2 + y_{13}V_3$$

$$I_2 = y_{21}V_1 + y_{22}V_2 + y_{23}V_3$$

$$I_3 = y_{31}V_1 y_{32}V_2 + y_{33}V_3$$

where

$$y_{11} = (Y_1 + Y_4 + Y_5),$$

$$y_{22} = (Y_2 + Y_5 + Y_6),$$

$$y_{33} = (Y_3 + Y_4 + Y_6),$$

$$y_{12} = y_{21} = -Y_5,$$

$$y_{13} = y_{31} = -Y_4, \text{ and}$$

$$y_{23} = y_{32} = -Y_6.$$

In matrix form,

$$\begin{bmatrix} I_1 \\ I_2 \\ I_3 \end{bmatrix} = \begin{bmatrix} y_{11} & y_{12} & y_{13} \\ y_{21} & y_{22} & y_{23} \\ y_{31} & y_{32} & y_{33} \end{bmatrix} \begin{bmatrix} V_1 \\ V_2 \\ V_3 \end{bmatrix}.$$

In general, the current vector, admittance matrix, and the voltage vector are related by

$$I_{BUS} = Y_{BUS} V_{BUS}$$

where the current in a bus is

$$I = \frac{S^*}{V^*} = \frac{P - jQ}{V^*}.$$

A systematic procedure may be used to develop the coefficient matrix of the matrix equation above. First, find out all the admittances that are connected to a node. Their sum is an element of the main diagonal. Don't change the sign of any admittance or the sum.

Secondly, find the admittance between a node and other node. The negative (or opposite in sign) of the admittance should be entered at the row and column of the nodes.

An example follows. From Figure , the admittances Y_1, Y_4, and Y_5 are all connected to node 1 of the figure. Take their sum and enter in element y_{11}. Next, from node 1 to node 2 is Y_5. Take its negative and enter in element y_{12}.

9.4 Example of a Two-Source, Three-Bus High-Voltage Transmission Network

Figure 9.3 is an example of a two-source, three-bus power transmission system. Busses 1 and 2 have both 1.0 per unit volt. All transmission lines have series inductive reactance of 0.1 ohm. There is no shunt capacitive reactance.

The equivalent admittance of each series inductive reactance is

$$Y = \frac{1}{j0.1} = -j10.$$

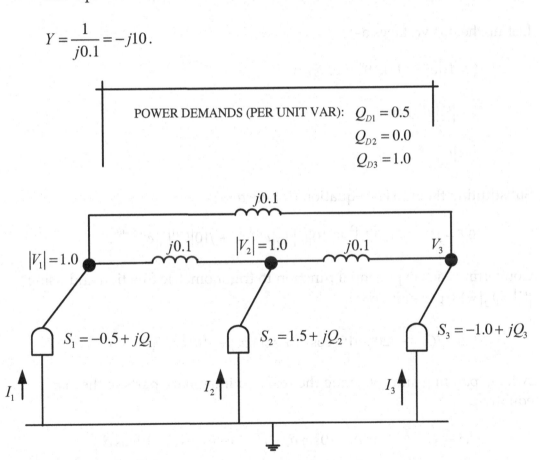

POWER DEMANDS (PER UNIT VAR): $\quad Q_{D1} = 0.5$
$Q_{D2} = 0.0$
$Q_{D3} = 1.0$

Figure 9.3 Example of a Two-Source, Three-Bus Power System

Since current is $I = \dfrac{S^*}{V^*} = \dfrac{P - jQ}{V^*}$ then

$$\dfrac{-0.5 - jQ_1}{V_1^*} = -j20V_1 + j10V_2 + j10V_3.$$

Multiplying both sides by V_1^* gives

$$-0.5 - jQ_1 = -j20V_1V_1^* + j10V_2V_1^* + j10V_3V_1^* = -j20|V_1|^2 + j10V_2V_1^* + j10V_3V_1^*.$$

Define the bus voltages as

$$V_1 = |V_1|e^{j\delta_1} = 1.0\angle 0^0 \text{ (i.e. } \delta_1 = 0^0),$$

$$V_2 = |V_2|e^{j\delta_2}, \text{ and}$$

$$V_3 = |V_3|e^{j\delta_3}.$$

Substituting them to last equation (for current),

$$-0.5 - jQ_1 = -j20|V_1|^2 + j10|V_1||V_2|e^{j(\delta_2 - 0)} + j10|V_1||V_3|e^{j(\delta_3 - 0)}.$$

Converting each exponential function to trigonometric function and using $|V_1| = |V_2| = 1.0$ per unit volt,

$$-0.5 - jQ_1 = -j20 + j10(\cos\delta_2 + j\sin\delta_2) + j10(\cos\delta_3 + j\sin\delta_3).$$

After expanding and collecting the real and imaginary parts of the last equation,

$$-0.5 - jQ_1 = -10\sin\delta_2 - 10\sin\delta_3 + j(-20 + 10\cos\delta_2 + 10\cos\delta_3).$$

Following the same procedure for bus 2

136

$$1.5 - jQ_2 = 10\sin\delta_2 - 10\sin(\delta_3 - \delta_2) + j(-20 + 10\cos\delta_2 + 10\cos(\delta_3 - \delta_2)),$$

and for bus 3,

$$-1.0 - jQ_3 = 10\sin\delta_3 - 10\sin(\delta_2 - \delta_3) + j(-20 + 10\cos\delta_3 + 10\cos(\delta_2 - \delta_3)).$$

The last three equations is a system of complex equations (or equations involving complex numbers). For the system to be true, the real parts on the left-hand side must equal the real part on the right-hand side. The same is true for their imaginary parts.

Collecting all the real parts,

$$-0.5 = -10\sin\delta_2 - 10\sin\delta_3,$$

$$1.5 = 10\sin\delta_2 - 10\sin(\delta_3 - \delta_2), \text{ and}$$

$$-1.0 = 10\sin\delta_2 - 10\sin(\delta_2 - \delta_3).$$

The collection of imaginary parts is:

$$-Q_1 = -20 + 10\cos\delta_2 + 10\cos\delta_3,$$

$$-Q_2 = -20 + 10\cos\delta_2 + 10\cos(\delta_3 - \delta_2), \text{ and}$$

$$-Q_3 = -20 + 10\cos\delta_3 + 10\cos(\delta_2 - \delta_3).$$

The collection of real-part equations has three unknowns. They are the power angles and their difference. If the unknowns can be found from the real-part equations, then all the reactive powers, from the collection of imaginary parts, can be found.

The problem with the above equations is the trigonometric function. It is not linear.

137

Consider two simultaneous equations involving the sine function. An example is

$$0.1 = \sin x + \sin y, \text{ and}$$

$$-0.5 = \sin y + \sin(x - y).$$

In order to find the solution of the system, first it is necessary to assume a value for x (in the first equation) and calculate for y. Next, the values of x and y are used in the second equation and see if the equality holds. If not, a new value for x is assumed and the process repeats until the second equation is satisfied. It is possible that no values for x and y can satisfy both equations. In this case, there is no solution.

9.4.1   Approximate solution to the two-source, three-bus example

An approximate solution to the two-source, three-bus problem may be found by assuming that for small angles,

$$\sin \delta = \delta \text{ radian.}$$

With this approximation, the first two equations of the real parts can be re-written as

$$-0.5 = -10\delta_2 - 10\delta_3$$

$$1.5 = 10\delta_2 - 10(\delta_3 - \delta_2).$$

The values of the power angles are

$$\delta_2 = 0.0667 \text{ radian, and}$$

$\delta_3 = -0.0167$.

Substituting the angles for the equation of the first reactive power,

$$-Q_1 = -20 + 10\cos\delta_2 + 10\cos\delta_3$$

gives

$Q_1 = 0.023$ per unit var.

For the remaining reactive powers,

$Q_2 = 0.057$ per unit var, and

$Q_3 = 0.036$ per unit var.

The reactive power demand of each bus is known (see Figure). Adding the reactive power in the line and the reactive power demand gives the generated reactive power. They are shown below:

$$Q_{G1} = Q_1 + Q_{D1} = 0.023 + 0.5 = 0.523 \text{ per unit var,}$$

$$Q_{G2} = Q_2 + Q_{D2} = 0.057 + 0.0 = 0.057 \text{ per unit var, and}$$

$$Q_{G3} = Q_3 + Q_{D3} = 0.036 + 1.0 = 1.036 \text{ per unit var.}$$

The total reactive power loss is the difference between the total generated (reactive) power and the total (reactive) demand. Since the total generated (reactive) power is 1.616 per unit var, and the total (reactive) demand is 1.5 per unit var, then the total (reactive) power loss is

$$Q_{LOSS} = Q_1 + Q_2 + Q_3 = \sum Q_{Gi} - \sum Q_{Di} = 1.616 - 1.5 = 0.116 \text{ per unit var.}$$

Since the resistance of the transmission lines is zero, the real power flow, from bus 1 to bus 3 is

$$P_{13} = \frac{|V_1||V_3|\sin(\delta_1 - \delta_3)}{X} \approx \frac{|V_1||V_3|(\delta_1 - \delta_3)}{X} = \frac{(1.0)(1.0)(0 - (-0.0167))}{0.1} = 0.167$$

per unit watt.

Similarly, the other power flows are:

$P_{12} = -0.667$ per unit watt, and

$P_{23} = 0.833$ per unit watt.

As for the reactive power flow, between bus 1 and bus 3,

$$Q_{13} = \frac{|V_1|^2 - |V_1||V_3|\cos(\delta_1 - \delta_3)}{X}.$$

Since for small angles, $\cos\delta = 1 - \dfrac{\delta^2}{2}$, the reactive power flow is

$Q_{13} = 0.001$ per unit var.

The other reactive power flows can be found similarly and they are:

$Q_{23} = 0.035$ per unit var, and

$Q_{12} = 0.022$ per unit var.

9.5 Gauss Iterative Technique in Solving High-Voltage Transmission Line Networks

Consider a function $f(x)$. Its solution is the value of x, such as $x = x_k$, that makes $f(x)$ zero. That is $f(x_k) = 0$.

Iterative technique of finding a solution is a guess-and-verify approach. First, it guesses an initial value for x. Next, it substitutes the value to x and verifies if the function is zero. If not, another a value is added or subtracted from the initial value and the process of verification is repeated again.

9.5.1 Basics of the Gauss iterative technique

In Gauss iterative technique, the function $f(x) = 0$ is modified to a form $x = F(x)$. Its solution is found when

$$x - F(x) = 0.$$

Suppose $f(x) = x^2 - 5x + 4$. Equating it to zero,

$$x^2 - 5x + 4 = 0.$$

Solving for x in $5x$,

$$x = \frac{x^2}{5} + \frac{4}{5}.$$

Assume the initial guess

$$x^{(0)} = 3.$$

The first iteration, using $x = \frac{x^2}{5} + \frac{4}{5}$, gives

$$x^{(1)} = \frac{3^2}{5} + \frac{4}{5} = 2.60.$$

For the second iteration, the calculated value in the first iteration is used. That is,

$$x^{(2)} = \frac{(2.60)^2}{5} + \frac{4}{5} = 2.152.$$

Note how the value of x is going down. If the equation has a solution, it will reach a steady state value such that x from the previous iteration is almost equal to the current iteration. At this time, the approximate solution is found.

9.5.2 Gauss iterative technique in solving problems in power transmission line networks

The iterative technique of Gauss can now be formalized as shown in the following paragraphs.

Define the current in a bus as

$$I_i = \frac{S_i^*}{V_i^*} = \frac{P_i - jQ_i}{V_i^*} = y_{ii}V_i + \sum_{\substack{k=1 \\ k \neq i}}^{n} y_{ik}V_k.$$

Its bus voltage is

$$V_i = \frac{1}{y_{ii}} \left(\frac{P_i - jQ_i}{V_i^*} - \sum_{\substack{k=1 \\ k \neq i}}^{n} y_{ik}V_k \right), \quad i = 2, 3, ..., n.$$

Note that, in its simplest form, the voltage is a function of voltage. That is,

$$V = f(V).$$

The Gauss technique can now be defined as

$$V_i^{(v+1)} = \frac{1}{y_{ii}}\left[\frac{P_i - jQ_i}{\left(V_i^{(v)}\right)^*} - \sum_{\substack{k=1 \\ k \neq i}}^{n} y_{ik}V_k^{(v)}\right], \ i = 2,3,...,n$$

where the superscript represents the iteration number.

Let

$$A_i \triangleq \frac{P_i - jQ_i}{y_{ii}}, \ i = 2,3,...,n$$

and

$$B_{ik} \triangleq \frac{y_{ik}}{y_{ii}}, \ i = 2,3,...,n, \ \text{and} \ k = 1,2,3,...,n \text{ but } k \neq i.$$

Then, Gauss iterative approach can be represented more compactly as

$$V_i^{(v+1)} = \frac{A_i}{\left(V_i^{(v)}\right)^*} - \sum_{\substack{k=1 \\ k \neq i}}^{n} B_{ik}V_k^{(v)}, \ i = 2,3,...,n \text{ and } k = 1,2,3,...,n.$$

At the first iteration, when $v = 0$, the voltage $V_i^{(0)}$ are assumed to be $1\angle 0^0$. The voltage $V_k^{(0)}$, however, is zero except for the swing bus, which is always $1\angle 0^0$ during the entire iteration.

In the second iteration, when $v = 1$, the voltage $V_i^{(1)}$ takes on the value calculated from the previous iteration. Similarly, the voltage, $V_k^{(1)}$ takes on the values calculated from the previous iteration.

9.5.3 Example of applying the Gauss iterative technique

Figure 9.4 is an example of a power transmission system for applying the iterative technique of Gauss. Before applying the technique, first obtain the Y_{BUS} matrix. The rules in generating the matrix are:

- For the main diagonal elements (driving point admittances), add all the admittances connected to a bus without changing the sign, and

- For the non-main-diagonal elements (transfer admittances), take the opposite (in sign) of the admittance.

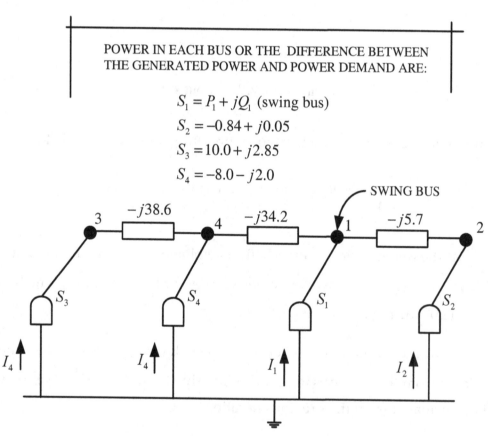

POWER IN EACH BUS OR THE DIFFERENCE BETWEEN
THE GENERATED POWER AND POWER DEMAND ARE:

$$S_1 = P_1 + jQ_1 \text{ (swing bus)}$$
$$S_2 = -0.84 + j0.05$$
$$S_3 = 10.0 + j2.85$$
$$S_4 = -8.0 - j2.0$$

Figure 9.4 A High-Voltage Transmission Network for Illustrating Gauss Iterative Technique

Using the rules, the admittance matrix is

$$Y_{BUS} = \begin{bmatrix} -j39.9 & j5.7 & 0 & j34.2 \\ j5.7 & -j5.7 & 0 & 0 \\ 0 & 0 & -j38.6 & j38.6 \\ j34.2 & 0 & j38.6 & -j72.8 \end{bmatrix}.$$

The next step is to calculate the constants

$$A_i \triangleq \frac{P_i - jQ_i}{y_{ii}}, \quad i = 2,3,...,n, \text{ and}$$

$$B_{ik} \triangleq \frac{y_{ik}}{y_{ii}}, \quad i = 2,3,...,n, \text{ and } k = 1,2,3,...,n \text{ but } k \neq i$$

With

$$S_1 = P_1 + jQ_1 \text{ (swing bus)},$$

$$S_2 = -0.84 + j0.05,$$

$$S_3 = 10.0 + j2.85, \text{ and}$$

$$S_4 = -8.0 - j2.0,$$

the coefficients are (A_1 not required):

$$A_2 = \frac{P_2 - jQ_2}{y_{22}} = \frac{-0.84 - (j0.05)}{-j5.7} = 0.1476\angle273.4°,$$

$$A_3 = \frac{P_3 - jQ_3}{y_{33}} = \frac{10.0 - (j2.85)}{-j38.6} = 0.2694\angle74.1°,$$

145

$$A_4 = \frac{P_4 - jQ_4}{y_{44}} = \frac{-8.0 - (-j2.0)}{-j72.8} = 0.1133 \angle 255.96^0,$$

$$B_{21} = \frac{y_{21}}{y_{22}} = \frac{j5.7}{-j5.7} = -1,$$

$$B_{23} = \frac{y_{23}}{y_{33}} = \frac{0}{-j5.7} = 0,$$

$$B_{24} = \frac{y_{24}}{y_{44}} = \frac{0}{-j72.8} = 0, \text{ and}$$

$$B_{32} = \frac{y_{32}}{y_{33}} = \frac{0}{-j38.6} = 0.$$

The third step to find the voltages by iteration using

$$V_i^{(v+1)} = \frac{A_i}{\left(V_i^{(v)}\right)^*} - \sum_{\substack{k=1 \\ k \neq i}}^{n} B_{ik} V_k^{(v)}, \quad i = 2, 3, ..., n \text{ and } k = 1, 2, 3, ..., n.$$

Iteration number 1:

$$V_2^{(1)} = \frac{A_2}{\left(V_2^{(0)}\right)^*} - \sum_{\substack{k=1 \\ k \neq i}}^{4} B_{2k} V_k^{(0)} = \frac{A_2}{\left(V_2^{(0)}\right)^*} - B_{21} V_1^{(0)} - B_{23} V_3^{(0)} - B_{24} V_4^{(0)}$$

$$= \frac{0.1476 \angle 273.41^0}{1 \angle 0^0} - \left(-1 \left(1 \angle 0^0\right)\right) - 0 - 0 = 1.0088 - j0.1473$$

$$V_3^{(1)} = \frac{A_3}{\left(V_3^{(0)}\right)^*} - \sum_{\substack{k=1 \\ k \neq i}}^{4} B_{3k} V_k^{(0)} = \frac{A_3}{\left(V_3^{(0)}\right)^*} - B_{31} V_1^{(0)} - B_{32} V_2^{(0)} - B_{34} V_4^{(0)}$$

$$= \frac{0.2694 \angle 74.1^0}{1 \angle 0^0} - 0 - 0 - \left(-1 \left(1 \angle 0^0\right)\right) = 1.1046 \angle 13.57^0$$

$$V_4^{(1)} = \frac{A_4}{\left(V_4^{(0)}\right)^*} - \sum_{\substack{k=1 \\ k \neq i}}^{4} B_{4k} V_k^{(0)} = \frac{A_4}{\left(V_4^{(0)}\right)^*} - B_{41} V_1^{(0)} - B_{42} V_2^{(0)} - B_{43} V_3^{(0)}$$

$$= \frac{0.1133 \angle 255.96^0}{1 \angle 0^0} - (-0.47)(1 \angle 0^0) - 0 - \left(-0.53\left(1 \angle 0^0\right)\right) = 0.9787 \angle -6.45^0$$

Iteration number 2 uses the voltages $V_2^{(1)}$, $V_3^{(1)}$, and $V_4^{(1)}$ to calculate the voltages. For bus 2,

$$V_2^{(2)} = \frac{A_2}{\left(V_2^{(1)}\right)^*} - \sum_{\substack{k=1 \\ k \neq i}}^{4} B_{2k} V_k^{(1)} = \frac{A_2}{\left(V_2^{(1)}\right)^*} - B_{21} V_1^{(1)} - B_{23} V_3^{(1)} - B_{24} V_4^{(1)}.$$

The same procedure should be followed for calculating the other bus voltages.

After the third iteration, the following values of the bus voltages are found:

$$V_2^{(3)} = 1.02 \angle -8.28^0,$$

$$V_3^{(3)} = 1.15 \angle 14.19^0, \text{ and}$$

$$V_4^{(3)} = 0.968 \angle -1.86^0.$$

Note that bus 1, being the swing bus, has the constant voltage of $V_1 = 1.0 \angle 0^0$.

The last step is finding the currents. In matrix form,

$$\begin{bmatrix} I_1 \\ I_2 \\ I_3 \\ I_4 \end{bmatrix} = \begin{bmatrix} -j39.9 & j5.7 & 0 & j34.2 \\ j5.7 & -j5.7 & 0 & 0 \\ 0 & 0 & -j38.6 & j38.6 \\ j34.2 & 0 & j38.6 & -j72.8 \end{bmatrix} \begin{bmatrix} V_1 \\ V_2 \\ V_3 \\ V_4 \end{bmatrix}$$

The calculated values of the voltages (after the third iteration) are used for the voltage vector.

As an example, for the current in the swing bus is

$$I_1 = y_{11}V_1 + y_{12}V_2 + y_{13}V_3 + y_{14}V_4$$
$$= (-j39.9)(1.0\angle 0^0) + (j5.7)(1.02\angle -8.28^0) + 0 + (j34.2)(0.968\angle -1.86^0)$$
$$= 1.9151 - j1.022$$

The conjugate of the power in the bus can be calculated using

$$I_1 = \frac{S^*}{V_1^*}.$$

Solving for S^*,

$$S^* = I_1 V_1^* = (1.9151 - j1.022)(1.0\angle 0^0) = 1.9151 - j1.022.$$

The bus power is, therefore,

$$S = (S^*)^* = 1.9151 + j1.022.$$

Since the generated power is the sum of the bus power and the power demand, the required amount of generated power can be found if the power demand is known.

Other currents may be found similarly. In addition, the power flow from a bus to any other bus can be found. If the resistance of the line is zero, then the real power flow from one bus to another bus can be found by taking the difference of all the real power coming into the bus and subtracting the real power demand. Note that this procedure applies only when the resistance is zero (and the real power loss is zero).

As far as the reactive power flow is concerned, the inductive reactance of a line is never zero. Hence, there will always be a reactive power loss. Therefore, the simple arithmetic of subtracting the reactive power demand from a bus will not work.

CHAPTER 10

FINANCING A POWER SYSTEM

Like any other business, financing a power system involves capital cost and operating cost. These costs determine the fuels to be used and the amount of power that can be allocated to each type of load. The chapter also presents the basics of optimum operating strategy to minimize the costs of operating power plants.

10.1 Types of Loads and Factors

The capacity factor of a power system is defined as

$$CF = \frac{P_A}{P_{MAX}}$$

where

CF = capacity factor,

P_A = actual power generated, and

P_{MAX} = maximum power available.

If a generator generates 120 MW from its maximum available power, then its capacity factor is $\frac{120}{200} = 0.60$ or 60%.

The load factor is the ratio of the average load and the maximum load. That is,

$$LF = \frac{L_A}{L_{MAX}}$$

Where,

LF = load factor,

L_A = average load, and

L_{MAX} = maximum load.

If the average load is 1,100 MW and the maximum load is 2,000 MW, then the load factor is $\dfrac{1,100}{2,000} = 0.55$ or 55%.

The capacity factor of a power system must match the load. Table 10.1 shows the types of loads and the capacity factor required to meet each of them. Figure 10.1 illustrates the location of the loads in graph called annual duration curve.

The annual duration curve is the cumulative sum of loads. Suppose in hour one the load is 1 MW and in hour two the load is 3 MW. Then, the cumulative load at hour two is 4 MW. That is, the present cumulative value is equal to the previous cumulative value plus the current value.

After 8,760 hours, the cumulative load represents the maximum or peak load. Dividing each cumulative value by the maximum or peak load gives the percent of each value. The graph of the annual duration curve may, therefore, be shown in terms of percent. In this sense, the annual duration curve is similar to cumulative probability function. The difference in the horizontal coordinates of two points on the curve represents the interval in hours. Its difference in the vertical coordinates, $|\Delta P|$ represents the probability of getting the interval from H to $H + \Delta H$ (see Figure 10.1).

Table 10.1 Types of Loads and their Capacity Factors

Load	Capacity Factor	Usual Source
Base	Greater than 60%	Coal, nuclear, or geothermal
Intermediate	30% to 60%	Oil, or natural gas
Peaking	Less than 30%	Gas turbine

10.2 Types of Costs

The linearized annual capital cost (LACC) converts a single large payment to a series of many small payments. It includes not only interest but also recovery costs. LACC is usually within the range of 15% to 25%.

Capital costs are installation costs per year. It is given by

$$CC = (C_E)(LACC)$$

where

CC = capital cost in $\dfrac{\$}{KW - year}$,

C_E = equipment cost in $\dfrac{\$}{KW}$, and

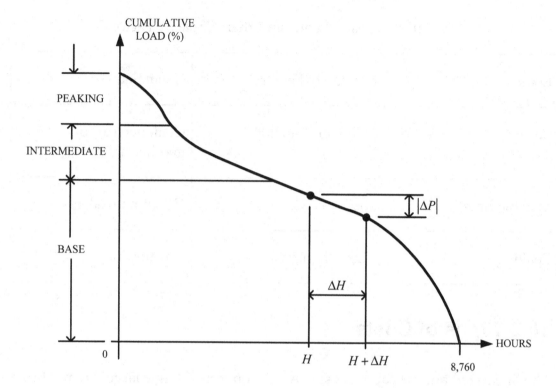

Figure 10.1 Annual Duration Curve Showing the Types of Loads

$LACC$ = linearized annual capital cost in $\dfrac{1}{year}$.

Operating cost has the same unit as the capital cost. It is defined as

$$OC = \left(C_F\right)\left(\frac{1}{Q_F}\right)(E)(H)$$

where

OC = operating cost in $\dfrac{\$}{KW - year}$,

C_F = cost of fuel in $\dfrac{\$}{unit}$,

Q_F = heat content of fuel in Btu,

E = efficiency in $\dfrac{Btu}{KWh}$, and

H = time the fuel is used in $\dfrac{hours}{year}$.

The efficiency of a fuel is the ratio of energy and its heat content. That is,

$$E = \frac{KWh}{Btu}.$$

Graphs of capital cost and operating cost are shown on Figure 10.2. The total cost is their sum. That is,

$$CT = CC + OC.$$

To obtain the total cost per year, multiply the total cost, in $\dfrac{\$}{KW - year}$, by the number of KW. That is,

$$C = (CT)(N_{KW})$$

where

C = total cost per year, and

N_{KW} = number of kilowatts produced in the year.

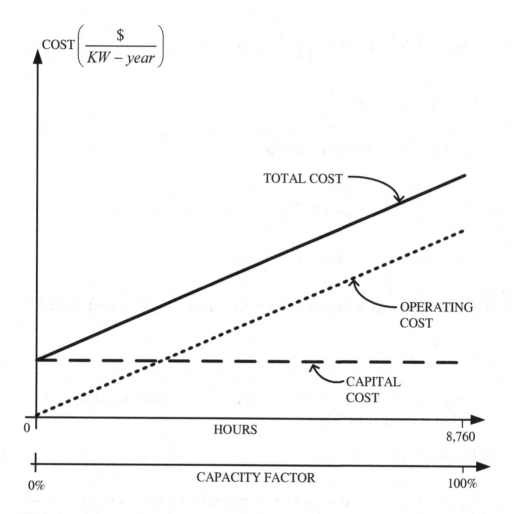

Figure 10.2 Graphs of Capital Cost, Operating Cost, and Total Costs

In power system operations, the capital cost does not vary. What varies is the operating cost.

10.3 Optimum Mix Design in Determining the Base, Intermediate, and Peaking Loads

Optimum mix design optimizes the allocation of fuel. The equations of the annual cost of fuels are plotted against the hours in a year. Points of intersections of two lines represent point of minimum costs. When these points are connected, the lines become optimum boundaries because cost below the line is guaranteed minimum.

Figure 10.3 shows the graphs of the annual cost of three fuels. Points A, B, and C represent the points. Projecting vertical lines from the points define the operating regions of the base load, the intermediate load, and the peaking load. Note that fuel #1 can't be used for the base load since its annual total cost is the highest (past point B). Similarly, fuel #2 can't be used for the base load either because its cost is higher than fuel #3 (past point B). Thus, the only choice for the base load is fuel #3. The same reasoning goes with the intermediate load, and the peaking load.

The base load, intermediate load, and the peaking load are determined by optimum mix design. Suppose that a power system will have a peak load of 4,000 MW. Table 10.2 shows the cost per kilowatt and capital cost of six different fuels that can be used. LACC is assumed to be 20%. Capital cost is the product of the cost per kilowatt and the LACC.

Table 10.3 shows the unit cost, heat content, and efficiency of each fuel type. The last column on the table is the operating cost. It is obtained by using the formula

$$OC = (C_F)\left(\frac{1}{Q_F}\right)(E)(H).$$

H in the formula is 8,8760 hours.

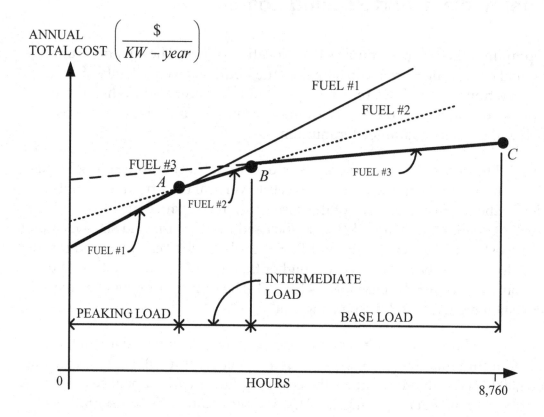

NOTE: ANNUAL TOTAL COST = ANNUAL
CAPITAL COST + OPERATING COST

Figure 10.3 Graphs of the Annual Total Costs of Three Different Fuels

The sum of the capital cost and the operating cost is shown on Table 10.4. To plot the equation of each cost, draw a line between the coordinates of the beginning point $(0, CC)$, and the coordinates of the ending point $(8760, TC)$. Figure 10.4 shows the graphs of the lines representing the total costs.

Table 10.2 Capital Costs of Different Fuels

Fuel	Cost ($/KW)	Capital Cost ($/KW-year)
Coal	1,200.00	240.00
Geothermal	1,700.00	340.00
Natural gas	500.00	100.00
Nuclear	1,500.00	300.00
Oil	700.00	140.00
Gas turbine	250.00	50.00

On the figure are two points A and B. Point A is the intersection of the lines representing gas turbine and natural gas. Left of the point is a solid representing gas turbine. This line segment represents the boundary of operating the peaking load.

Point B is the intersection of natural gas and coal. Between points A and B is a solid line segment representing the operating boundary of the

Table 10.3 Operating Costs of the Fuels

Fuel	Unit Cost, C_F in $/unit	Heat Content, Q_F in Btu/unit	Efficiency, E in Btu/KWh	Operating Cost, OC in $/KW-year
Coal	45.00/ton	21,000,000/ton	9,500	173.37
Geothermal	160/million pounds	211/pound	31,000	205.92
Natural gas	6.74/1,000 cubic feet	1,050/cubic feet	10,200	573.55
Nuclear	44/pound	26,400,000/pound	10,800	157.68
Oil	40/barrel	6,200,00/barrel	10,100	570.81
Gas turbine	42/barrel	5,800,000/barrel	13,000	824.00

intermediate load. To the right of point B is the last line segment representing coal. This is the boundary for the base load.

Points A and B are projected onto the horizontal axis of the annual duration curve shown on Figure 10.5. They are then projected further to the curve itself. Next, their intersections with the curves are projected (horizontally) onto the vertical axis of the annual duration curve. The

Table 10.4 Capital Costs, Operating Costs and Total Costs of the Fuels

Fuel	Capital Cost, CC ($/KW-year)	Operating Cost, OC ($/KW-year)	Total Cost, TC ($/KW-year)
Coal	240.00	173.37	413.37
Geothermal	340.00	205.92	545.92
Natural gas	100.00	573.55	673.55
Nuclear	300.00	157.68	457.68
Oil	140.00	570.81	710.81
Gas turbine	50.00	824.00	874.00

projections divide the vertical axis into three regions (base load, intermediate load, and the peaking load).

Suppose the peak or maximum load is 4,000 MW. The annual duration curve has also examples of the percentages between the loads. From the percentages, the amount of loads for each types are:

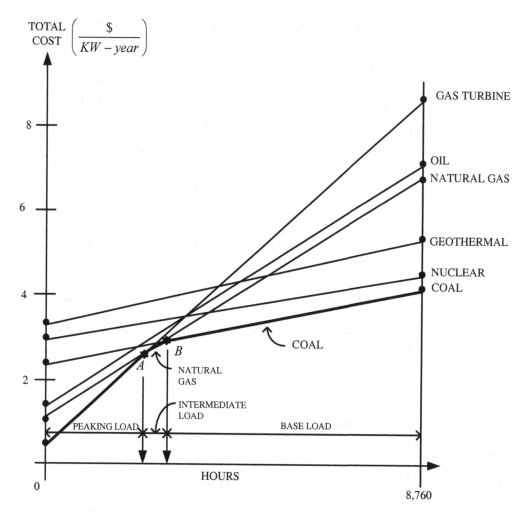

Figure 10.4 Plots of the Total Costs of Fuels

$$P_{PEAKING} = (100\% - 72\%)(4,000) = 1,120 \text{ MW},$$

$$P_{INTERMEDIATE} = (72\% - 70\%)(4,000) = 80 \text{ MW, and}$$

$$P_{BASE} = (70\% - 0)(4{,}000) = 2{,}800 \text{ MW}.$$

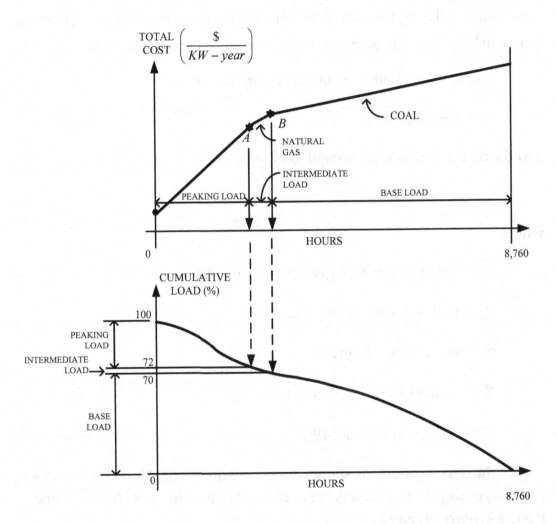

Figure 10.5 Determining the Base, Intermediate, and the Peaking Loads

10.4 Optimum Operating Strategy

Suppose a power plant has generator, G_1 and another power plant has generator, G_2. The optimum power dispatch or operating strategy determines how much power each generator should produce to minimize cost.

Define the cost function of the first generator as

$$C_1 = \alpha_1 + \beta_1 P_{G1} + \lambda_1 P_{G1}{}^2,$$

and the cost function of the second generator as

$$C_2 = \alpha_2 + \beta_2 P_{G2} + \lambda_2 P_{G2}{}^2$$

where

C_i = cost of operating a generator in \$/hour,

P_{Gi} = power generated by generator i,

α_i = constant in \$/hour,

β_i = constant in \$/MW-hour, and

λ_i = constant in $/ hour(MW)^2$.

The cost functions, as shown above, are quadratic. Figure shows their (generic) examples. In practice, curve fitting can be employed to determine the constants α, β, and λ.

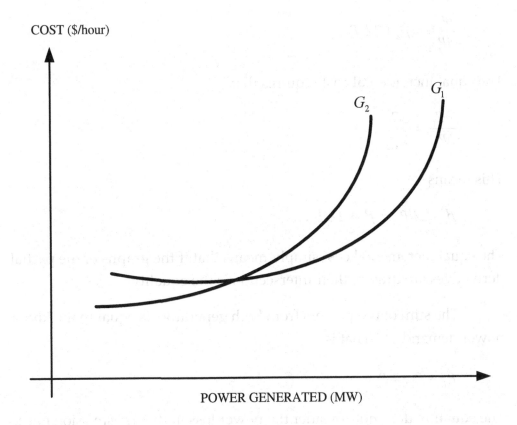

COST ($/hour)

G_2

G_1

POWER GENERATED (MW)

Figure 10.6 Cost Functions of Generators

Taking the partial derivatives of each cost function

$$\frac{\partial C_1}{\partial P_{G1}} = \beta_1 + 2\lambda_1 P_{G1},$$

and

$$\frac{\partial C_2}{\partial P_{G2}} = \beta_2 + 2\lambda_2 P_{G2}.$$

The equal incremental cost requires that

$$\frac{\partial C_1}{\partial P_{G1}} = \frac{\partial C_2}{\partial P_{G2}}.$$

This means

$$\beta_1 + 2\lambda_1 P_{G1} = \beta_2 + 2\lambda_2 P_{G2}.$$

The equal incremental cost simply means that if the graphs of the partial derivatives are drawn, their intersection is the equality.

The sum of the powers from both generators is equal to the known power demand, P_D. That is,

$$P_{G1} + P_{G2} = P_D$$

The equation does not consider the power loss in the transmission lines.

The only unknowns, from the last two equations are P_{G1} and P_{G2}. Solving for them,

$$P_{G1} = \left(\frac{\lambda_2}{\lambda_1 + \lambda_2}\right) P_D + \frac{1}{2}\left(\frac{\beta_2 - \beta_1}{\lambda_1 + \lambda_2}\right),$$

and

$$P_{G2} = \left(\frac{\lambda_1}{\lambda_1 + \lambda_2}\right) P_D + \frac{1}{2}\left(\frac{\beta_1 - \beta_2}{\lambda_1 + \lambda_2}\right).$$

Notice that the above equations are equations of a straight line. The coefficient of the first term is the slope of the line. Its last term is the intercept at the vertical axis.

The above formulas are based on the assumption that the losses in the transmission lines are zero. Hence, the generated power, as calculated from the formulas, will be less. To include the effect of the losses in the transmission lines, replace the factor, P_D by $(P_D + P_L)$ where P_L is the sum of all the losses. The procedure is justified since the power demand is (assumed) constant and so is the sum of the losses. That is, if an equation has a constant, the same equation can be used if the same constant is replaced by another constant.

Power from each generator is now given by

$$P_{G1} = \left(\frac{\lambda_2}{\lambda_1 + \lambda_2} \right)(P_D + P_L) + \frac{1}{2} \left(\frac{\beta_2 - \beta_1}{\lambda_1 + \lambda_2} \right),$$

and

$$P_{G2} = \left(\frac{\lambda_1}{\lambda_1 + \lambda_2} \right)(P_D + P_L) + \frac{1}{2} \left(\frac{\beta_1 - \beta_2}{\lambda_1 + \lambda_2} \right).$$

The last issue is when three or more plants are supplying power to the demands and the losses. To solve this problem, use the method of the optimum mix design.

First, get all the partial derivatives of all the generators. They all should describe a straight line. Plot all the lines. Next, examine the intersection of the lines. The points of intersection represent an optimum strategy. Identify the equations of the two lines intersecting at a point and equate the two. Repeat the procedure for all the points. The procedure should give the amount of power that each generator should produce.

CHAPTER 11

A DESIGN PROJECT

This chapter shows an example of a design project. After showing the broad outline of the design procedure, details are shown. The example uses a seven-bus transmission network. An iterative approach is used in finding its solution. It is not the intent of the project to show the result of the design as the most economical or the most reliable. Rather, it is showing the procedure. Several trials are required to obtain the most economical and reliable result.

11.1 Outline of the Design Procedure

The design of a power system consists of the following steps:

- Perform optimum mix design to select the fuels to be used,

- Analyze the real power (not reactive power) of the busses,

- Propose the configuration of the lines and the selection of transformers and protective devices,

- Calculate impedances and admittances, and

- Perform the iterative technique on the matrices to find voltages, currents, generated power, power flow, and power losses.

    As can be seen from the procedure, the design of a high-voltage power transmission system is highly mathematical. Compounding the problem is the lack of uniqueness in the solution. Hence, a trial and error process is used in the design.

11.1.1   Optimum mix design

The method of optimum mix design is shown in chapter 10. There are other factors that must be considered in optimum mix design. They are governmental regulations and costs. Design must comply with these constraints.

11.1.2   Real power (not reactive power) analysis of the busses

In the real power analysis of a bus, the real power loss in a line is neglected. That is, the resistance of the line is assumed zero. Note that the real power flow is the sum of the real power loss plus the real power that adjacent bus will need. The calculations of the real power in a bus should determine how much real power that bus should generate.

11.1.3   Select the number of circuits, configuration of the tower, transformers, and protective devices

Once the amount of real power is known in each bus, the next step is to select the number of circuits, configuration of transmission line towers, transformers, and protective devices (or circuit breakers). Two factors affect the selection. They are the economics of the selection and the N-1 outage. In the N-1 outage, a transmission line corridor has N lines. N-1 outage means one of the lines is cut reducing N lines to N-1 lines. Essentially, the N-1 constraint ensures that the system can still deliver power to all demands when one of the lines is cut.

11.1.4   Calculations of the impedances and admittances

Usually, the design of a high-voltage network has no more than 100-mile bus-to-bus maximum distance. Part of the network may have a 20-mile distance.

Therefore, the usual case is to treat the transmission lines as either a short line or a medium length transmission line. Calculate the impedances of the lines according chapter 6. Separate calculations should be made for cases with N-1 outages.

Shunt admittances may be calculated using the formulas developed in chapter 7. In some cases, approximate values of the shunt admittance may be given by a utility company.

11.1.5   Perform the iterative approach to solve the matrix equation

The presentation of the iterative technique of Gauss, chapter 9, is sufficiently complete. It may be transformed to an algorithm and then to a computer program. Otherwise, the designer may use commercially available software. If the latter is chosen, ensure that the results of its calculations are accurate. Use examples of networks in textbooks and simulate them using the software. Needless to say, the software must not be used if it is not accurate.

11.2 Example of a Design Project

Figure 11.1 shows a 7-bus power transmission system for the project. New lines and generators are to be installed in the system.

Except for bus 1 and 7, the projected peak demand on each bus is shown as a downward arrow. There are two existing 300-MW oil-fired power plants in bus 6. Constraints on a bus, such as the maximum power or the type of fuel, are shown on a bus. Figure 11.2 sows the bus-to-bus distances. The impedance and admittance of an existing line can be assumed to be equal to that of new lines.

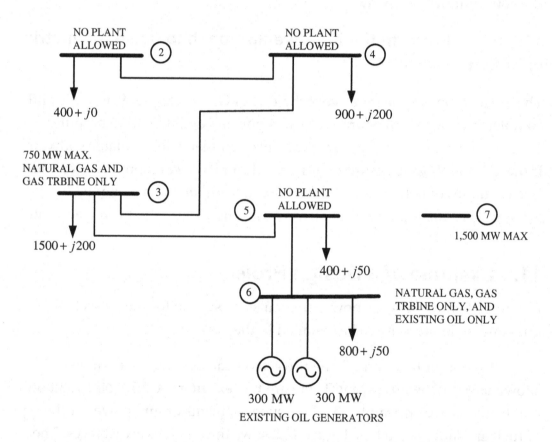

Figure 11.1 Peak Demands in the Busses of the Design Project

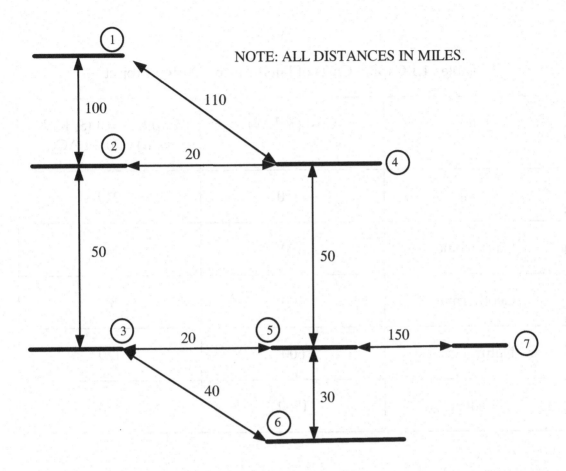

Figure 11.2 Distances of the Busses in the Design Project

Optimum Mix Design

Tables 11.1 to 11.3 show the costs of fuels for the power plants, their calculated annual capital costs, operating costs, and total costs. Using the method developed in chapter 10, the graphs of the total costs, for each type of fuel, are shown on Figure 11.3. Points A and B are the two critical points on the graphs. Vertical projections from these points, onto the annual duration

Table 11.1 Capital Costs of Fuels for the Design Project

Fuel	Cost ($/KW)	Capital Cost ($/KW-year) @20% LACC
Coal	1100	220
Gas turbine	400	80
Geothermal	400	80
Natural gas	600	120
Nuclear	1500	300

curve (not shown), defines the base load, intermediate load, and the peaking load. The results of such projections show that the mix of the fuels are distributed as follows:

- Base load (coal) = 50% of 4,000 MW = 2,000 MW,

- Intermediate load (natural gas) = 27.1% of 4,000 MW = 1,083 MW, and

- Peaking load (geothermal and existing oil) = 22.9% of 4,000 MW = 917 MW.

Table 11.2 Operating Costs of the Fuels for the Design Project

Fuel	Unit Cost, C_F in $/unit	Heat Content, Q_F in Btu/unit	Efficiency, E in Btu/KWh	Operating Cost, OC in $/KW-year
Coal	$30/ton	21,600,000 Btu/ton	9,500	115.58
Gas turbine	$40/barrel	5,800,000 Btu/barrel	14,500	876.00
Geothermal	$530/million pounds	211 Btu/pound	31,000	682.11
Natural gas	$3.32/1,000 cubic feet	1,050 Btu/cubic feet	10,200	282.52
Nuclear	$44/pound	26,400,000 Btu/pound	10,800	157.68

At this point it is instructive to find the nominal size of a plant for each type of fuel. Table 11.4 shows the nominal sizes of power plants with a particular fuel type.

A 10% reserve, above the total peak demand of 4,000 MW, is required. This means that that the system must have a total capacity of 4,400 MW.

Table 11.3 Capital Costs and Total Costs of Power Plants for the Design Project

Plant with Fuel	Capital Cost ($/KW-year)	Total Cost ($/KW-year)
Coal	220	335.58
Gas turbine	80	956.00
Geothermal	80	762.12
Natural gas	120	402.52
Nuclear	300	457.68

Power from the two existing oil generators may be used as part of the reserve.

Inspection of the constraints placed on the design of the system shows that of the four generator sites, only two sites can have coal. They are bus 1 and bus 7. Furthermore, the only possible location for the geothermal plant is bus 7. Therefore, power plants, using natural gas, can only be located at bus 3 and bus 6.

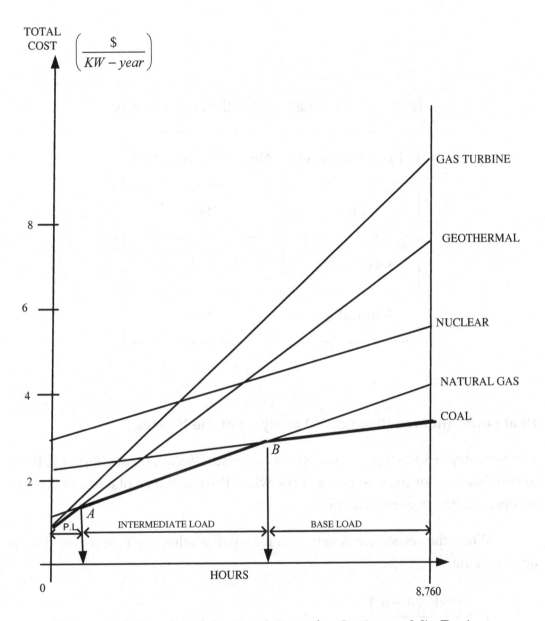

Figure 11.3 Graphs of the Total Costs for Optimum Mix Design

Table 11.4 Nominal Sizes of the Power Plants

Plant with Fuel	Nominal Size (MW)
Coal	1000
Geothermal	100
Natural gas	300

Real power (not reactive power) analysis of the busses

The next step is to assign power plants to busses. Recall, from chapter 9, the approximation for the real power flow when the resistance of a line is zero. It is repeated here for convenience.

When the resistance is zero and for small angles, the real power flow is approximately

$$P_{ij} \approx \frac{|V_i||V_j|(\delta_i - \delta_j)}{X}.$$

Smaller real power flow decreases the voltage magnitudes. More importantly, the power angle or the difference in the angles of the two busses decreases. These are favorable since they increase the stability of the system.

To minimize the real power flow, the real power of all generators (directly) feeding a bus must be equal to the peak demand. If there is transmission line between the bus and the demand bus, that transmission line may be regarded as non-existing.

Since the total demand in bus 2 and bus 3 is 1,300 MW, then a coal (with 1,000-MW nominal size) and a natural gas (with 300-MW nominal size) must feed bus 1.

The remaining peak demand of 2,700 MW must be satisfied by power plants in bus 7, bus 3, and bus 6. Bus 7 must use the remaining 1,000 MW base load from coal. Additionally, since bus 7 is the only location for the geothermal plant, bus 7 must also have the five 100-MW geothermal plants for peaking load.

With 1,500 MW feeding 2,700 MW, the remaining peak demand (to match) is 1,200 MW. Add to this the reserve requirement of 400 MW and the result is 1,600 MW. Since the two oil generators has total capacity of 600 MW, then the number of natural gas to use is 1,000 MW. Each natural gas has a nominal size of 300 MW. Therefore, the number of required natural gas plants is four. Their total capacity is 1,200 MW. It is more than the 1,000 MW remaining demand (peak and reserve). The extra 200 MW

The total peak demand in bus 6 is 800 MW. Since the operating cost of the existing oil generators is higher than that of the natural gas, the number of natural gas generators in the bus must be two. It can't be one since it will maximize the operation of the oil generators. Additionally, can't be three either since it will exceed the 800-MW peak demand.

With two of the four natural gas power plants sited at bus 6, the remaining two natural gas plants must be at bus 3.

Figure 11.4 shows the assignments of the power plants. The figure also shows the operating power of each plant.

Selection of the circuits, transformers, and circuit breakers

The design also requires at least two transmission lines leaving a bus with 500 MW or more of peak demand. It must also withstand N-1 transmission line outage. Table 11.5 shows the possible circuits and voltage of the lines. Other equipment, such as transformers and circuit breakers, are shown on Tables 11.6 and 11.7.

The two constraints above control the type of circuits that can be used. For example, one 500-KV line can't replace two 230-KV lines feeding a 500-MW bus. That is, when the 500-KV line is out the system failed the N-1 outage. Furthermore, the design requires at least two lines.

Given the requirements, the best approach is to use the 230-KV line. Whenever two circuits are required, the vertical configuration costing $500,000 per mile will be used. Otherwise, if a single circuit is required, the horizontal configuration will be used. The latter costs $300,000 per mile. Each circuit has a thermal capacity of 440 MVA.

Starting at bus 4, the bus has 900 MW peak demand. Since a circuit has 440-MVA capacity, two lines are required from bus 1 to bus 4. The existing line from bus 2 to bus 4; and, the existing line from bus 5 to bus 4 should provide the additional load. If one of the two lines is cut, the remaining line will not be able to provide the power. A line has thermal capacity of 440

MVA. With a power factor of 0.80, the 440 MVA is equivalent to 352 MW. Hence, one 230-KV line will not be sufficient from bus 1 to bus 2. Two lines

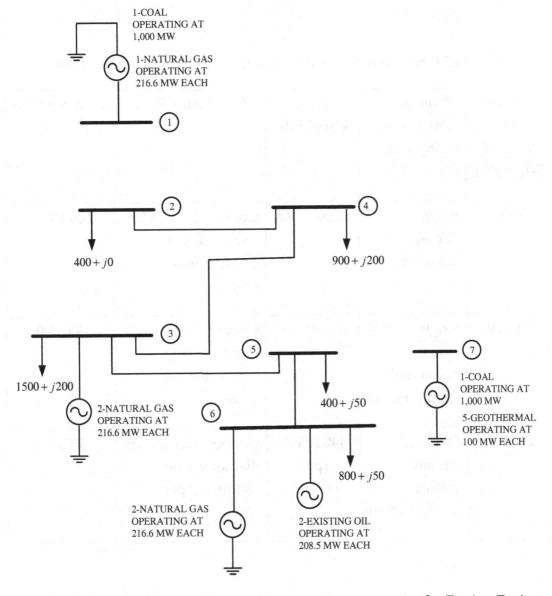

Figure 11.4 Assignments of Power Plants or Generators in the Design Project

Table 11.5 Characteristics of the Transmission Lines and their Costs

Line-to-Line Voltage (Kilovolts)	Number of Circuits and configuration	Load Capability	Conductor Type	Cost per Mile ($/mile)
500 KV	Single circuit, horizontal	1,200 MVA	Copper, 1.75 inches diameter, two conductors per phase	400,000
230 KV	Single circuit, horizontal configuration	440 MVA	Copper, 1.21 inches diameter, one conductor per phase	300,000
230 KV	Double circuit, vertical configuration	440 MVA (per circuit)	Copper, 1.21 inches diameter, one conductor per phase	500,000

Table 11.6 Characteristics of the Transformers and their Costs

Line-to-Line Voltage (Kilovolts)	Rating (MVA)	Impedance (% on own base)	Conductor Type	Cost
500/230 KV	500 MVA	j10	Copper, 1.75 inches diameter, two conductors per phase	$6 million
500/230 KV	1,200 MVA	j10	Copper, 1.21 inches diameter, one conductor per phase	$9 million

Table 11.7 Costs of the Circuit Breakers

Line-to-Line Voltage	Cost
500 KV	$1 million each
230 KV	$400,000 each

are required. If a line, from bus 1 to bus 4, is cut off, the power from bus 1 to bus 2 will flow to bus 4 via the existing line from bus 2 to bus 4.

The next bus to analyze is bus 3. It has 1,500 MW of peak demand and total generating capacity of 433.2 MW. Taking the difference gives approximately 1,067 MW required power flow from bus 7. Since the load at bus 5 will take 400 MW from the 1,500 MW at bus 7, the available power to bus 3 is approximately 1,100 MW. The 1,067 MW load can be matched by the 1,100 MW available power. Dividing the 1,067 MW by 352 MW gives 3.0 lines. However, there is an existing line from bus 4 to bus 3 and another existing line from bus 5 to bus 3. Therefore, the number of new lines reduces from 3.0 to 2.0. If one of the lines is cut, the remaining lines should be able to match the peak demand.

Bus 6 has 800 MW of peak demand. It can be matched by the generators. No circuit is required.

The last bus is bus 7. It has to deliver 1,500 MW to bus 5. It requires 1,500/352 = 4.26 lines or 5 lines. It is directly feeding a 400-MW load and the constraint that it must satisfy the N-1 outage is not required.

Figure11.5 is the layout of the lines. They are represented as heavy black lines with squares as circuit breakers at the ends. The lines and busses are rated at 230 KV.

The design requires that all distribution loads are fed from 230-KV bus. Since, all the busses in the circuits are 230 KV, no transformer is required.

Calculations of the series impedances and shunt admittances

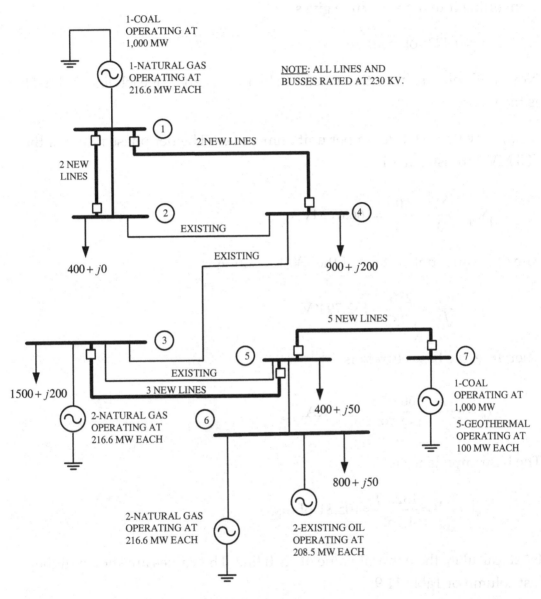

Figure 11.5 Layout of the Transmission Lines (230-KV) in the Design Project

The calculated series inductance of the 230-KV line is 0.00048 ohm per meter. Converting it to ohm per mile gives

$$X = 0.7726 \, \text{ohm/mile.}$$

Next, multiply the inductive reactance by the distance of each line. Table 11.8 is the result.

Next, convert the values in per unit ohm values. The per phase rating of the 500 MVA transformer is

$$|S_{1\phi}| = \frac{|S_{3\phi}|}{3} = \frac{500}{3} = 166.67 \, \text{MVA.}$$

The line-to-neutral voltage of the 230-KV rating is

$$V_{LN} = \frac{V_{LL}}{\sqrt{3}} = \frac{230}{1.732} = 132.79 \, \text{KV.}$$

Therefore, the base current is

$$I_B = \frac{|S_{1\phi}|}{V_{LN}} = \frac{166.67}{132.79} = 1.255 \, \text{KA.}$$

The base impedance is

$$|Z_B| = \frac{V_{LN}}{I_B} = \frac{132.79}{1.255} = 105.81 \, \text{ohms.}$$

Next, calculate the per unit value of each line. The values are shown in the last column of Table 11.9.

The admittance of the transmission line, as provided by a utility company, is approximately

Table 11.8 Inductive Reactance of the 230-KV Line (in Ohm)

Between Busses	Length (miles)	Inductive Reactance (ohms)	Number of Parallel Circuits	Inductive Reactance of Parallel Circuits (ohms)
1 and 2	100	77.2485	2	38.6242
1 and 4	110	84.933	2	42.4866
2 and 4	20	15.4497	1	15.4497
4 and 5	50	38.6243	1	38.6242
3 and 5	20	15.4497	4	15.4497
7 and 5	150	115.8727	5	23.1745
5 and 6	30	23.1745	1	23.1745

$\dfrac{Y}{2} = 0.008$ per unit mho per mile.

Multiplying the rate with distance of a line gives the admittance for the line. When two circuits are in parallel, their equivalent admittance is twice the single circuit. Calculated equivalent admittance of the lines are shown on Table 11.10.

Table 11.9 Inductive Reactance of the 230-KV Line (in Per Unit Ohm)

Between Busses	Length (miles)	Number of Parallel Circuits	Inductive Reactance of Parallel Circuits (ohms)	Per Unit Ohm of the Parallel Circuits
1 and 2	100	2	38.6242	0.365
1 and 4	110	2	42.4866	0.401
2 and 4	20	1	15.4497	0.146
4 and 5	50	1	38.6242	0.365
3 and 5	20	4	15.4497	0.146
7 and 5	150	5	23.1745	0.219
5 and 6	30	1	23.1745	0.219

Iterative approach to solve the matrix equation

The last step is using an iterative approach similar to Gauss' iterative technique. Due to the immense number of calculations (and required accuracy), manual calculation is not recommended.

In addition to the series impedances and shunt admittances of the lines, software using the approach will also ask for the number of busses,

Table 11.10 Admittances of the Lines Using 230 KV

Between Busses	Length (miles)	Number of Parallel Circuits	Equivalent Y/2 of the Parallel Circuits (Per Unit Mho)
1 and 2	100	2	1.6
1 and 4	110	2	1.76
2 and 4	20	1	0.16
4 and 5	50	1	0.40
3 and 5	20	4	0.64
7 and 5	150	5	6.0
5 and 6	30	1	0.24

number of lines, voltage control bus, the swing bus, line numbers, starting and ending busses of a line, acceleration factor (for the calculations), the minimum and maximum limits for the reactive power, complex power of the (peak) demands, and the real power of each generator.

The software will repeat the same calculations until it converge to a solution. If the calculations do not converge then the system is unstable. If it converged, the software will give bus voltages, complex power flows, currents, and losses in the system.

Tables 11.11 and 11.12 are samples of the results. The first table, or Table 11.11, shows the complex voltage of each bus and its complex power.

Notice that no bus has an angle close to 45 degrees. It indicates the system, as designed, is stable.

Table 11.12 shows the complex power flow between two busses. The direction of the real power flow, from bus i to bus j, is always opposite of the direction from bus j to bus i. In contrast, the reactive power flow has no such symmetry. Review section 9.4. In the section, the collection of equations, with real parts only, is used to find the bus angles and their differences. In contrast, the collection of equations, with imaginary parts only, is used to find the reactive power flows. Each reactive power flow must be unique.

Table 11.11 Results of the Iterative Approach for Bus Voltages and Power

Bus	Voltage Magnitude (per unit volt)	Voltage Angle (degrees)	Real Power of the Bus (MW)	Reactive Power of the Bus (MVAR)
1	1.000	0.0	1196.0	919.9
2	0.800	-32.83	-400.0	0.0
3	1.000	-37.00	-1,067.0	760.8
4	0.786	-38.01	-900.0	-200.0
5	0.948	-32.31	-400.0	-50.0
6	1.000	-30.93	51.0	120.1
7	1.000	11.62	1,500.0	719.2

Table 11.12 Results of the Iterative Approach for Power Flows

Line Number	Sending Bus	Ending Bus	Real Power Flow (MW)	Reactive Power Flow (MVAR)
1	1	2	593.956	447.676
1	2	1	-593.956	-45.418
2	1	4	602.978	472.196
2	4	1	-602.978	-2.706
3	2	4	194.591	45.849
3	4	2	-194.591	-27.80
4	5	4	101.391	214.156
4	4	5	-101.391	-168.992
5	5	3	1060.310	-637.729
5	3	5	-1060.310	760.924
6	7	5	1501.102	719.214
6	5	7	-1501.102	486.706
7	6	5	52.113	120.085
7	5	6	-52.113	-113.010

REFERENCES

1. Abramowitz, Milton, and Stegun, Irene A., editors, Handbook of Mathematical Functions, 9th printing, New York: Dover Publications, 1965.

2. Chen, Wai-Kai, Editor-in chief, The Circuits and Filters Handbook, Massachusetts, CRC Press, Inc., 1995.

3. De Sosa, Jesus, Design, Analysis, and Maintenance of Electrical and Electronic Systems in Facilities, iUniverse, 2010.

4. De Sosa, Jesus, Power, Testing, and Grounding of Electronic Systems, Indiana: iUniverse, 2008.

5. De Sosa, Jesus, Relays and Instrument Transformers in Protecting Power Systems, iUniverse, 2010.

6. Kiameh, Philip, Power Generation Handbook, Selection, Applications, Operation, and Maintenance, New York: McGraw-Hill, 2003.

7. Kreyzig, Erwin, Advanced Engineering Mathematics, New York: John Wiley & Sons, Inc., 1979.

8. Stevenson, William D., Elements of Power System Analysis, McGraw Hill, 4th ed., 1982.

9. Van Valkenberg, M. E., Network Analysis, 3rd edition, New Jersey: Prentice-Hall, 1974.